Population Genomics in the Developing World

Translational and Applied Genomics Series
Population Genomics in the Developing World
Concepts, Applications, and Challenges

Series editor

George P. Patrinos

Laboratory of Pharmacogenomics and Individualized Therapy, Department of Pharmacy, School of Health Sciences, University of Patras, Patras, Greece; Department of Genetics and Genomics, College of Medicine and Health Sciences, United Arab Emirates University, Al-Ain, Abu Dhabi, United Arab Emirates; Zayed Center for Health Sciences, United Arab Emirates University, Al-Ain, Abu Dhabi, United Arab Emirates; Clinical Bioinformatics Unit, Department of Pathology, Faculty of Medicine and Health Sciences, Erasmus University Medical Center, Rotterdam, the Netherlands

Edited by

Marlo Möller

South African Medical Research Council Centre for Tuberculosis Research, Division of Molecular Biology and Human Genetics, Department of Biomedical Sciences, Stellenbosch University, Stellenbosch, South Africa; Centre for Bioinformatics and Computational Biology, Stellenbosch University, Stellenbosch, South Africa

Caitlin Uren

South African Medical Research Council Centre for Tuberculosis Research, Division of Molecular Biology and Human Genetics, Department of Biomedical Sciences, Stellenbosch University, Stellenbosch, South Africa; Centre for Bioinformatics and Computational Biology, Stellenbosch University, Stellenbosch, South Africa

ACADEMIC PRESS

An imprint of Elsevier

Academic Press is an imprint of Elsevier
125 London Wall, London EC2Y 5AS, United Kingdom
525 B Street, Suite 1650, San Diego, CA 92101, United States
50 Hampshire Street, 5th Floor, Cambridge, MA 02139, United States

Notices
Knowledge and best practice in this field are constantly changing. As new research and experience broaden our understanding, changes in research methods, professional practices, or medical treatment may become necessary.

Practitioners and researchers must always rely on their own experience and knowledge in evaluating and using any information, methods, compounds, or experiments described herein. In using such information or methods they should be mindful of their own safety and the safety of others, including parties for whom they have a professional responsibility.

To the fullest extent of the law, neither the Publisher nor the authors, contributors, or editors, assume any liability for any injury and/or damage to persons or property as a matter of products liability, negligence or otherwise, or from any use or operation of any methods, products, instructions, or ideas contained in the material herein.

ISBN: 978-0-443-18546-5

For information on all Academic Press publications visit our website at https://www.elsevier.com/books-and-journals

Publisher: Mica Haley
Acquisitions Editor: Ali Afzal-Khan
Editorial Project Manager: Susan E. Ikeda
Production Project Manager: Jayadivya Saiprasad
Cover Designer: Greg Harris

Typeset by TNQ Technologies

Working together to grow libraries in developing countries

www.elsevier.com • www.bookaid.org

Contents

CHAPTER 9   Direct-to-consumer (DTC) genetic testing and the
population genomics industry **159**
Caitlin Uren and Desiree C. Petersen

Contributors

Analabha Basu
National Institute of Biomedical Genomics, Kalyani, West Bengal, India

Nadia Carstens
Division of Human Genetics, National Health Laboratory Service and School of Pathology, Faculty of Health Sciences, University of the Witwatersrand, Johannesburg, South Africa; South African Medical Research Council, Genomics Platform, Cape Town, South Africa

Emile R. Chimusa
Department of Applied Sciences, Faculty of Health and Life Sciences, Northumbria University, Newcastle, Tyne and Wear, United Kingdom

Ananyo Choudhury
Sydney Brenner Institute for Molecular Bioscience, Faculty of Health Sciences, University of the Witwatersrand, Johannesburg, South Africa

Dayna A. Croock
South African Medical Research Council Centre for Tuberculosis Research, Division of Molecular Biology and Human Genetics, Faculty of Medicine and Health Sciences, Stellenbosch University, Cape Town, South Africa

Katelyn Cuttler
South African Medical Research Council Centre for Tuberculosis Research, Division of Molecular Biology and Human Genetics, Faculty of Medicine and Health Sciences, Stellenbosch University, Cape Town, South Africa

Nicole van der Merwe
Division of Chemical Pathology, Department of Pathology, Faculty of Medicine and Health Sciences, Stellenbosch University, Cape Town, South Africa; FamGen Counselling, Bloemfontein, South Africa

Brigitte Glanzmann
South African Medical Research Council Centre for Tuberculosis Research, Division of Molecular Biology and Human Genetics, Faculty of Medicine and Health Sciences, Stellenbosch University, Cape Town, South Africa; South African Medical Research Council, Genomics Platform, Cape Town, South Africa

Kariofyllis Karamperis
Laboratory of Pharmacogenomics and Individualized Therapy, Department of Pharmacy, School of Health Sciences, University of Patras, Patras, Greece; Group of Algorithms for Population Genomics, Department of Genetics, Institut de Biologia Evolutiva, IBE, (CSIC-Universitat Pompeu Fabra), Barcelona, Spain; The Golden Helix Foundation, London, United Kingdom

Craig Kinnear
South African Medical Research Council Centre for Tuberculosis Research, Division of
Molecular Biology and Human Genetics, Faculty of Medicine and Health Sciences,
Stellenbosch University, Cape Town, South Africa; South African Medical Research Council,
Genomics Platform, Cape Town, South Africa

Maritha J. Kotze
Faculty of Medicine and Health Sciences, Stellenbosch University and National Health
Laboratory Service, Tygerberg Hospital, Division of Chemical Pathology, Department of
Pathology, Cape Town, South Africa

Elouise Elizabeth Kroon
South African Medical Research Council Centre for Tuberculosis Research, Division of
Molecular Biology and Human Genetics, Faculty of Medicine and Health Sciences,
Stellenbosch University, Cape Town, South Africa

Amanda J. Lea
Department of Biological Sciences, Vanderbilt University, Nashville, TN, United States

Amica Corda Müller-Nedebock
South African Medical Research Council Centre for Tuberculosis Research, Division of
Molecular Biology and Human Genetics, Faculty of Medicine and Health Sciences,
Stellenbosch University, Cape Town, South Africa; Stellenbosch University, South African
Medical Research Council/Stellenbosch University Genomics of Brain Disorders Research Unit,
Cape Town, South Africa

Marlo Möller
South African Medical Research Council Centre for Tuberculosis Research, Division of
Molecular Biology and Human Genetics, Faculty of Medicine and Health Sciences,
Stellenbosch University, Cape Town, South Africa; Centre for Bioinformatics and Computational
Biology, Stellenbosch University, Cape Town, South Africa

Maria A. Nieves-Colón
Department of Anthropology, University of Minnesota Twin Cities, Minneapolis, MN, United
States

George P. Patrinos
Laboratory of Pharmacogenomics and Individualized Therapy, Department of Pharmacy,
School of Health Sciences, University of Patras, Patras, Greece; Department of Genetics and
Genomics, College of Medicine and Health Sciences, United Arab Emirates University, Al-Ain,
Abu Dhabi, United Arab Emirates; Zayed Center for Health Sciences, United Arab Emirates
University, Al-Ain, Abu Dhabi, United Arab Emirates; Clinical Bioinformatics Unit, Department
of Pathology, Faculty of Medicine and Health Sciences, Erasmus University Medical Center,
Rotterdam, the Netherlands

Desiree C. Petersen
South African Medical Research Council Centre for Tuberculosis Research, Division of
Molecular Biology and Human Genetics, Faculty of Medicine and Health Sciences,
Stellenbosch University, Cape Town, South Africa

Michèle Ramsay
Sydney Brenner Institute for Molecular Bioscience, Faculty of Health Sciences, University of the Witwatersrand, Johannesburg, South Africa

Austin W. Reynolds
Department of Microbiology, Immunology, and Genetics, School of Biomedical Sciences, University of North Texas Health Science Center, Fort Worth, TX, United States

Dhriti Sengupta
Sydney Brenner Institute for Molecular Bioscience, Faculty of Health Sciences, University of the Witwatersrand, Johannesburg, South Africa

Carene Anne Alene Ndong Sima
South African Medical Research Council Centre for Tuberculosis Research, Division of Molecular Biology and Human Genetics, Faculty of Medicine and Health Sciences, Stellenbosch University, Cape Town, South Africa

Anwani Siwada
South African Medical Research Council Centre for Tuberculosis Research, Division of Molecular Biology and Human Genetics, Faculty of Medicine and Health Sciences, Stellenbosch University, Cape Town, South Africa

Yolandi Swart
South African Medical Research Council Centre for Tuberculosis Research, Division of Molecular Biology and Human Genetics, Faculty of Medicine and Health Sciences, Stellenbosch University, Cape Town, South Africa

Nchangwi Syntia Munung
Division of Human Genetics, Department of Pathology. Faculty of Health Sciences, University of Cape Town, Cape Town, South Africa

Caitlin Uren
South African Medical Research Council Centre for Tuberculosis Research, Division of Molecular Biology and Human Genetics, Department of Biomedical Sciences, Stellenbosch University, Stellenbosch, South Africa; Centre for Bioinformatics and Computational Biology, Stellenbosch University, Stellenbosch, South Africa

Roland van Rensburg
Division of Clinical Pharmacology, Department of Medicine, Faculty of Medicine & Health Sciences, Cape Town, South Africa

Population genomics—The fundamentals

Desiree C. Petersen and Dayna A. Croock

South African Medical Research Council Centre for Tuberculosis Research, Division of Molecular Biology and Human Genetics, Faculty of Medicine and Health Sciences, Stellenbosch University, Cape Town, South Africa

A list of terms used in this chapter is found in Box 1.1.

Box 1.1 Glossary

Population: *Organisms of the same species that occupy the same space at the same time and have the capability of interbreeding.*

Population genetics: *The study of genetic variation within and among populations and the evolutionary factors that explain this variation, including nonrandom mating, mutation, gene flow, genetic drift, and natural selection.*

Gene: *The unit of heredity that is passed on from one generation to the next. It is a combination of four different nucleotides (A, G, C, T) that form the genetic sequence of an individual. The gene encodes a functional molecule, often a polypeptide. The order of nucleotides determines the order of amino acid molecules of the polypeptide.*

Allele: *Alternative forms of a gene (arising by mutation of a single nucleotide) that are found at the same location on a chromosome and control the same trait.*

Locus: *Location where an allele is found on a chromosome.*

Gene pool: *Collection of all genes within an interbreeding population. It is the sum of all the alleles at all the loci within the genes of a population of a single species.*

Genotype: *Genetic constitution of an organism and refers to the alleles carried in its DNA, which give rise to the phenotype or observable traits of an organism.*

Phenotype: *Observable traits of an organism resulting from the interaction of the genotype and the environment.*

Hardy-Weinberg Equilibrium: *An important fundamental principle of population genetics, which states that the allele and genotype frequencies in a population will remain constant from generation to generation in the absence of other evolutionary influences.*

Homozygous: *Having two identical alleles of a particular gene or genes.*

Heterozygous: *Having two different alleles of a particular gene or genes.*

Dominant: *One allele of a gene on a chromosome masking or overriding the effect of the second allele on the other copy of the chromosome.*

Recessive: *An allele that does not create a phenotype if a dominant allele is present.*

Mutation: *An alteration in the nucleic acid sequence of the genome.*

Gene flow: *Any movement of individuals, and/or the genetic material they carry, from one population to another (also called migration).*

Genetic drift: *The change in frequency of an existing gene variant in the population due to random chance.*

Natural selection: *The process whereby organisms better adapted to their environment tend to survive and produce more offspring. The theory of its action was first fully expounded by Charles Darwin, and it is now regarded as the main process that brings about evolution.*

Population stratification: *Differences in allele frequencies between cases and controls due to systematic differences in ancestry rather than association of genes with disease.*

Population Genomics in the Developing World. https://doi.org/10.1016/B978-0-443-18546-5.00001-2

The Hardy-Weinberg principle

The Hardy-Weinberg principle is the cornerstone of population genetics.[1] Following the rediscovery of Gregor Mendel's work in 1900, opponents of the Mendelian view of inheritance contended that dominant traits should increase in frequency in a population over time, which is not observed in genuine populations. In 1908, British mathematician G. H. Hardy Fig. 1.1 refuted this argument in his paper "Mendelian proportions in a mixed population," in which he demonstrated that the allele frequencies in an infinitely large, random-mating population will not change from generation to generation.[2] The principle was known as Hardy's Law until 1943 when it was realized that German physician Wilhelm Weinberg Fig. 1.1 independently derived the same principle a few months before Hardy. However, Weinberg's paper was published in German and thus remained unknown to most English-speaking geneticists for 35 years.[3,4]

The Hardy-Weinberg theorem states that the frequencies of the genotypes obtained from two alleles remain constant through generations. Consider a population with two alleles (the dominant A allele and the recessive a allele) that segregate at a single locus, as per the laws of Mendelian inheritance. The frequency of allele A is denoted by p and the frequency of allele a is denoted by q. It then follows that the frequency of the homozygous dominant genotype AA is p^2 and the frequency of the homozygous recessive genotype aa is q^2. The frequency of the heterozygous genotype (which can be denoted as Aa or aA) will be $2pq$ Fig. 1.2.

FIGURE 1.1 G. H. Hardy (left) and Wilhelm Weinberg (right).

Hardy and Weinberg's theorems contributed to the Hardy-Weinberg principle.

From Edwards AWF. G. H. Hardy (1908) and Hardy-Weinberg equilibrium. Genetics. 2008;179(3):1143–1150. https://doi.org/10.1534/genetics.104.92940.

Gametes		Male	
		A (*p*)	a (*q*)
Female	A (*p*)	AA (p^2)	Aa (*pq*)
	a (*q*)	Aa (*pq*)	aa (q^2)

FIGURE 1.2 Hardy-Weinberg equilibrium genotypes.

A Punnett square depicting the probabilities of generating all possible genotypes at a biallelic locus in a population that conforms to Hardy-Weinberg assumptions.

If *A* and *a* are the only two alleles available at that locus, then $(p + q)$ equals one. Accordingly, the Hardy-Weinberg genotype frequencies—the binomial expansions of $(p + q)^2$—will also equal one.

$$p^2 + 2pq + q^2 = 1 \tag{1.1}$$

It is possible to apply the Hardy-Weinberg theorem to loci with more than two alleles with small changes to the Hardy-Weinberg equation. For example, for a locus with three possible alleles, $(p + q + r)^2$ will equal one.

This equation permits the prediction of allele and genotype frequencies for populations in Hardy-Weinberg equilibrium Box 1.2.

Box 1.2 Example 1

A farm has a population of black and white cats. The allele for black coat color (*A*) has complete dominance over the allele for white coat color (*a*) Fig. 1.3. Given a population of 1000 cats (840 black cats and 160 white cats), determine the allele frequencies and the number of cats with *AA*, *Aa*, and *aa* genotypes.

Step 1:

Determine the proportion of white cats in the population. These cats have only one possible genotype (*aa*). Thus the frequency of white cats in the population will be equal to the frequency of the *aa* genotype, which is denoted as q^2 in the Hardy-Weinberg equation.

$$\frac{160}{1000} = 0.16 \tag{1.2}$$

Therefore $q^2 = 0.16$.

Step 2:

Find *q* by the square root of q^2.

Continued

Box 1.2 Example 1—cont'd

$$\sqrt{q^2} = \sqrt{0.16} \tag{1.3}$$

$$q = 0.4 \tag{1.4}$$

Step 3:
Determine p from $p + q = 1$

$$p = 0.6 \tag{1.5}$$

Step 4:
Solve for the genotype frequencies using the Hardy-Weinberg equation $p^2 + 2pq + q^2 = 1$.

$$p = 0.6 \tag{1.6}$$

$$p^2 = 0.36 \tag{1.7}$$

$$2pq = 2(0.6)(0.4) \tag{1.8}$$

$$2pq = 0.48 \tag{1.9}$$

Therefore:
The frequency of the dominant alleles: $p = 0.6$.
The frequency of the recessive alleles: $q = 0.4$.
The frequency of cats with the dominant genotype: $p^2 = 0.36$.
The frequency of cats with the heterozygous genotype: $2pq = 0.48$.
The frequency of cats with the recessive genotype: $q^2 = 0.16$.

	Gametes	Male	
		A (p)	a (q)
Female	A (p)	AA (p^2) 	Aa (pq) 
	a (q)	Aa (pq) 	aa (q^2) 

FIGURE 1.3 A Punnett square depicting the probabilities of generating all possible genotypes in the black and white cat population.

The dominant allele is denoted by *A*. The recessive allele is denoted by *a*.

However, the conclusions of the Hardy-Weinberg theorem and equation apply only when the population conforms to the following assumptions:

1. Mating between individuals in the population must be random.
2. The mutation rate is negligible (i.e., mutations must not occur to introduce new alleles to the population).
3. There must be no migration (gene flow) between populations that will increase variability in the gene pool.
4. The population size is infinitely large to ensure allele frequencies are not changed through genetic drift.
5. There is no natural selection to alter the allele frequencies within the population.

If the genotype frequencies in a population deviate from Hardy-Weinberg expectations (due to inbreeding, for example), only one generation of random mating is required to return the genotype frequencies to equilibrium proportions. However, if the allele frequencies in a population change due to an external evolutionary disturbance (for example, genetic drift), the population will reach a new equilibrium state after a single generation of random mating. Thereafter a population that meets Hardy-Weinberg assumptions will remain at the new equilibrium until perturbed again.

One can ascertain if a population is in Hardy-Weinberg equilibrium by comparing the expected genotype frequencies (predicted from the Hardy-Weinberg equation) with the observed genotype frequencies in the population Box 1.3.

Box 1.3 Example 2

The genetic variability in a population of African penguins is being investigated. As part of this research, the genotype frequencies in this population are calculated to determine if the population is in Hardy-Weinberg equilibrium.

Researchers obtained the genotype distributions at one locus for 64 penguins Table 1.1:

Step 1:

Calculate the allele frequencies:

Each penguin with the SS genotype has two copies of the S allele.

Each penguin with the heterozygous genotype has one copy of the S allele and one copy of the F allele.

Each penguin with the FF genotype has two copies of the F allele Table 1.2.

To calculate the allele frequencies, divide the number of S or F alleles by the total number of alleles (128 alleles).

S: $p = 94/128 = 0.734$.

F: $q = 34/128 = 0.266$.

Step 2:

Calculate the expected genotype frequencies using the Hardy-Weinberg equation.

$p^2 = (0.734)^2 = 0.539$.

$2pq = 2(0.734)(0.266) = 0.390$.

$q^2 = 0.071$.

Step 3:

Calculate the number of penguins expected to have each genotype based on the Hardy-Weinberg equation by multiplying the number of individuals genotyped (n = 64) with the expected genotype frequencies obtained in step 2 Table 1.3.

Step 4:

Perform a chi-squared test to determine if the observed and expected genotype frequencies are significantly different from each other. The chi-squared test can be done using online statistical software and even Microsoft Excel. A

Continued

Box 1.3 Example 2—cont'd

chi-squared test can determine if the observed (O) and expected (E) genotype frequencies differ significantly using a chi-squared test.[5]

$$x^2 = \sum \frac{(O - E)^2}{E}$$

A population is **not** in Hardy-Weinberg equilibrium when $X^2 \geq 3.841$ and $p \leq 0.05$.

In our penguin population, $X^2 = 2.495$, $p = 0.2872$. Thus the observed and expected genotype frequencies do not differ significantly. This population is in Hardy-Weinberg equilibrium.

Table 1.1 Genotype distributions in a population of penguins.

Genotype	Frequency
SS	37
SF	20
FF	7

Table 1.2 Allele frequencies in a population of penguins.

Genotype	Observed frequency	Alleles
SS	37	74 S alleles
SF	20	20 S alleles 20 F alleles
FF	7	14 F alleles

Table 1.3 Expected and observed allele frequencies in a population of penguins.

Genotype	Observed frequency	Expected frequency
SS	37	34.496
SF	20	24.960
FF	7	4.544

Why is the Hardy-Weinberg theorem important?
Evolutionary implications of the Hardy-Weinberg theorem

The Hardy-Weinberg theorem is a null model for population genetics and evolution. Populations that meet the assumptions for Hardy-Weinberg equilibrium are said to be non-evolving because their allele frequencies are not changing over time. As mentioned previously, one can use the Hardy-Weinberg equation to predict allele and genotype frequencies in a population. However, real populations frequently violate the Hardy-Weinberg assumptions. First, populations are finite and are thus susceptible to genetic drift. Additionally, some forms of natural selection (e.g., balancing selection) can preserve Hardy-Weinberg equilibrium proportions. Hardy-Weinberg equilibrium can also appear to be maintained in populations that have very low rates of migration or mutation undetectable by current statistical methods.

Quality control in genetic association studies

Deviations in Hardy-Weinberg equilibrium proportions can also indicate the presence of genotyping errors and undetected population stratification. Chi-squared goodness-of-fit tests are performed on single nucleotide polymorphism (SNP) data to identify sites that deviate from Hardy-Weinberg equilibrium. These SNPs can be filtered out of the dataset to control for type 1 errors.

Genetic counseling

Genetic counselors can use allele and genotype frequency estimations from the Hardy-Weinberg principle to better quantify risk for individuals who do not have a family history of the disease but are part of a population group with a high prevalence of the inherited condition.

The Hardy-Weinberg theorem tells us about allele frequencies within a population and how these remain constant over generations. The next section will describe patterns of allele frequencies between populations.

Evolutionary forces on populations

There are five factors known to influence evolution due to changes in allele frequencies within a population. This may also lead to changes in various physical traits Fig. 1.4. These factors, often driven by environmental changes, include nonrandom mating, mutation, gene flow, genetic drift, and natural selection. Evolutionary factors can occur together and may collectively contribute to constant changes in genetic variation in the gene pool of a population. Therefore when evolutionary influences are present, the specific conditions required to maintain Hardy-Weinberg equilibrium are not kept and there is potential for population evolution.[6]

Nonrandom mating

Nonrandom mating occurs when the selection of a mate is influenced by the presence of specific physical traits or the preference to mate with others that are more similar to themselves. This selection

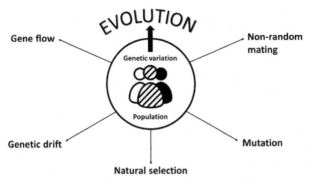

FIGURE 1.4 Five factors contributing to evolutionary influences on populations.

Gene flow, nonrandom mating, genetic drift, natural selection, and mutation.

is therefore influenced by observable phenotypes, which are determined by the underlying genotype. There are two types of nonrandom mating: inbreeding and assortative mating. Both of these nonrandom mating types in human populations are largely dependent on physical characteristics, ethnicity, religion, culture, and geographical location.[7–9]

Inbreeding

Inbreeding is a type of mating that takes place between closely related individuals that share genetic relatedness. These are termed consanguineous matings, involving one or more common ancestors, and their offspring therefore have an increased probability of receiving two copies of a specific allele. This results in identical copies of a single allele inherited from the common ancestor. Inbreeding at a population level results in altering genotype frequencies, more specifically, an increase in homozygous genotypes and a decrease in heterozygous genotypes, although it does not affect allele frequencies Box 1.4. This could lead to a decrease in overall fitness with many inherited autosomal recessive disorders being observed more often in populations that display high consanguinity rates Fig. 1.5.[10]

Box 1.4 Consanguinity and Hardy-Weinberg Allele Frequencies

In section The Hardy-Weinberg principle of this chapter, we discussed the Hardy-Weinberg equation. Example 1 demonstrated the use of the equation to determine if a population is in Hardy-Weinberg equilibrium.

$$p^2 + 2pq + q^2 = 1$$

Populations in Hardy-Weinberg equilibrium will not experience a change in allele frequencies from one generation to the next. This rule also applies to populations where consanguineous matings occur. The offspring of consanguineous matings have an increased probability of receiving two copies of a specific allele. This may result in an increased proportion of homozygous individuals in the population and a decrease in the proportion of heterozygous individuals. This mathematically satisfies the Hardy-Weinberg equation, which will always equal one.

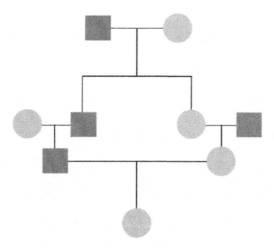

FIGURE 1.5 Pedigree analysis showing consanguineous mating between first cousins.

Pink circles represent females and blue squares represent males.

Assortative mating

Assortative mating refers to mating between those with similar phenotypes or among those found in a particular location. Mating therefore occurs among those more genetically similar to each other than would be expected by chance with two randomly chosen mates. Examples of assortative mating in humans include favorable phenotypic traits, such as height and weight influenced by specific genotypes, as well as the consideration of population stratification due to social groupings and geographic proximity. Assortative mating also results in the increase of the relative frequency of homozygous genotypes that could lead to a decrease in overall fitness; however, the effect is less than observed for inbreeding. As with inbreeding, assortative mating also does not change allele frequencies within a population Fig. 1.6.[11]

FIGURE 1.6 Assortative mating.

Mating between genetically similar individuals found in the same location.

Mutation

A mutation refers to specific changes in the DNA sequence, which could result in changing the frequency of various alleles that contribute to genetic variation in a population. The mutation of a single base present in more than 1% of the population is defined as a SNP. The mutation rate is low for most species and they are passed on to subsequent generations with subtle effects on allele frequencies over time between generations. Although mutations could result in changing the genotypes present in a population, this may or may not directly influence the observation of different phenotypes. Mutations may impact genes or entire chromosomes; however, they typically occur as either point mutations or base-pair insertions/deletions. With many mutations being harmful, they can reduce lifespan or the ability to reproduce, although they tend to be selected against and removed from the population with time.[12] If a mutation has an advantage over a wild-type allele, it will likely persist and spread among populations. The most documented example showing selective advantage of a mutation is heterozygotes carrying the sickle cell allele at the hemoglobin-beta gene (*β-globin*) locus that is highly protective against malaria caused by *Plasmodium falciparum*. While individuals with the normal homozygous genotype remain susceptible to malaria, it is the homozygotes for the sickle cell allele that are prone to severe anemia and early death due to the sickling of red blood cells. Therefore of the three genotypes, it is the advantageous heterozygous state that offers the highest level of fitness Fig. 1.7.[13]

Gene flow

Gene flow is also known as gene migration and occurs when allele frequencies change in a population due to transfer of alleles into or out of distinct gene pools. Mating could take place between adjacent populations resulting in genotype frequencies being altered and genetic variation being increased or decreased in the gene pool. Gene flow is therefore dependent on migration between separated

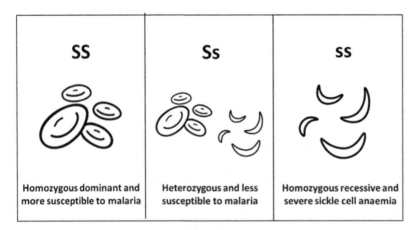

SS	Ss	ss
Homozygous dominant and more susceptible to malaria	Heterozygous and less susceptible to malaria	Homozygous recessive and severe sickle cell anaemia

FIGURE 1.7 Sickle cell disease is a homozygous dominant disorder resulting in sickling of red blood cells.
Heterozygous advantage is displayed by carriers of the sickle cell allele (*Ss*) who are highly protected against malaria.

populations with the possibility of adding new alleles to a population and thereby changing the composition by increasing the genetic variation in the original gene pool. Gene migration out of a population can also alter allele frequencies with the removal or decrease in specific alleles occurring within the gene pool resulting in a loss of genetic diversity Fig. 1.8.[14]

Genetic drift

Genetic drift involves a change in the allele frequencies of a population that occurs by chance and not by adaptation due to natural selection. The alleles that remain may therefore be either helpful or harmful to the population. The impact of genetic drift is greater when the population is smaller as there is a higher chance of some alleles becoming either fixed or extinct. When considering larger populations, allele frequencies are more likely to be maintained due to the occurrence of specific alleles in a large number of the population. The removal of alleles from a population will change both the allele and genotype frequencies and there are specific bottleneck events that promote genetic drift resulting in lower genetic diversity within a population.[15]

Bottleneck effect

When a population is sharply reduced in size by a natural disaster, it is termed a bottleneck effect Fig. 1.9. This often occurs due to a catastrophic event that eradicates most of the population with specific allele frequencies drifting toward becoming either increased or decreased. The surviving population therefore displays limited genetic diversity represented by these alleles and there is a reduced gene pool to draw from. It has been suggested that population bottlenecks are responsible for the reduced genetic diversity observed in modern human populations following migration out of Africa. A bottleneck effect within an existing population may result in a founder effect.[16]

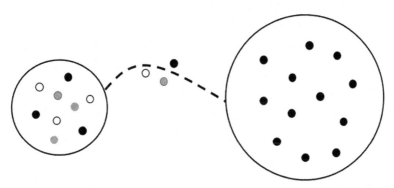

FIGURE 1.8 Gene flow.

Variation from the gene pool of one population is introduced into another population through migration.

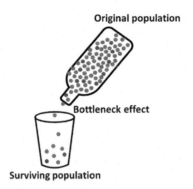

FIGURE 1.9 Bottleneck effect.

The bottleneck results in reduced genetic diversity of alleles in the surviving population compared to the original population.

Founder effect

When a small group becomes separated from the original population to form a new population, it is termed a founder effect Fig. 1.10. The new population therefore does not have the full representation of alleles found in the original population, which results in different allele frequencies being observed in

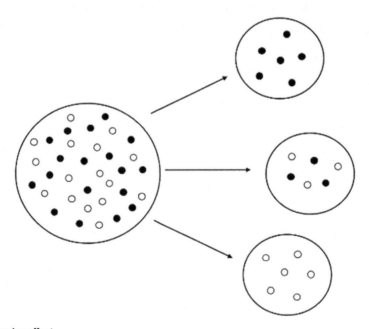

FIGURE 1.10 Founder effect.

The three new populations have different allele frequencies to each other and to the original founder population.

the comparatively smaller gene pool. Examples of founder effects include the presence of specific genetic disorders with a higher prevalence in isolated populations, such as variegate porphyria and familial hypercholesterolemia, observed in the South African Afrikaner population.[17]

Natural selection

Natural selection is the process whereby populations that are better adapted to their environment tend to survive and consequently produce more offspring (positive selection). It occurs when one allele or a combination of alleles across different genes (known as polygenic) influences traits and increases fitness. Natural selection therefore acts on phenotype thereby changing the genotype of the gene pool with more favorable alleles being passed on within the population. Unlike genetic drift, natural selection changes the allele frequencies in a population not due to chance, but rather due to environmental adaptation.[18] An example of natural selection is the presence of a genetic mutation for lactose tolerance in human populations where agriculture was introduced. Although the majority of the world's population is lactose intolerant, a point mutation involving a change from the cytosine (C) allele to the thymine (T) allele in the minichromosome maintenance complex component 6 (*MCM6*) gene will increase the fitness of adult humans by allowing the digestion of milk and milk products.[19]

While negative selection acts against deleterious mutations and positive selection acts on favorable mutations,[20] balancing selection maintains multiple alleles at a gene locus to increase genetic diversity (e.g., heterozygote advantage).[21] To ensure the survival of a population, it is vital to maintain genetic variation and select new traits with beneficial phenotypes. Natural selection serves as an important factor in driving evolution and there are three different types of selection based on phenotypic traits that can be observed, namely stabilizing selection, directional selection, and disruptive selection.[18,22]

Stabilizing selection

Stabilizing selection (Fig. 1.11) is considered to be the most common type of natural selection whereby a population displays intermediate phenotypes as having the highest fitness, while extreme phenotypes are selected against. This type of selection therefore over time removes extreme phenotypes and favors the narrowing of observed phenotypes in a population, which could alter allele frequencies and a subsequent decrease in genetic diversity. The standard distribution of phenotypes is therefore observed during stabilizing selection.

FIGURE 1.11 Stabilizing selection.

The intermediate phenotype is more successful and extreme phenotypes are selected against.

FIGURE 1.12 Directional selection.

The distribution of phenotypes changes over time from one extreme phenotype to another extreme phenotype.

FIGURE 1.13 Disruptive selection.

Two extreme phenotypes are more successful than the intermediate phenotype.

Directional selection

Directional selection (Fig. 1.12) represents a type of natural selection where one of the extreme phenotypes have the highest fitness in a population. This will also result in the allele frequency changing over time. The distribution is therefore shifted toward one side for selection of this extreme and more fit phenotype.

Disruptive selection

Disruptive selection (Fig. 1.13) is not as common and is a type of natural selection where the two extreme phenotypes are considered having the highest fitness compared to the intermediate phenotype. The allele frequencies representing these selected extreme phenotypes will therefore increase over time. The distribution is shown as two higher peaks on either end that represent the extreme phenotypes with selection against the intermediate phenotype.

Linkage disequilibrium, haplotypes, and allele frequency patterns
Meiosis and recombination

Haploid gametes are produced by the process of meiosis, a specialized type of cell division that reduces the number of chromosomes in a cell by half.[23] Meiosis is an essential feature of sexual reproduction and contributes to genetic diversity in offspring through the processes of recombination and independent assortment.

During the first phase of meiosis, the genetic material replicates and homologous pairs of chromosomes align along the equator of the cell. During this alignment, the arms of the homologous chromosomes overlap and temporarily fuse. Crossing over results in the exchange of genetic material between the homologous chromosomes, a process known as recombination Fig. 1.14.

Homologous chromosomes contain the same genes, but the maternal and paternal chromosomes in a homologous pair may have different forms of the same gene (termed alleles). Recombination creates new combinations of alleles, producing gametes that are genetically unique to the parent cells.

Genes that are located a distance away from each other on a chromosome are more likely to be separated during homologous recombination. However, genes that are located close to each other on a chromosome are less likely to be separated during recombination and are thus likely to be inherited together. This phenomenon is known as genetic linkage Fig. 1.15.

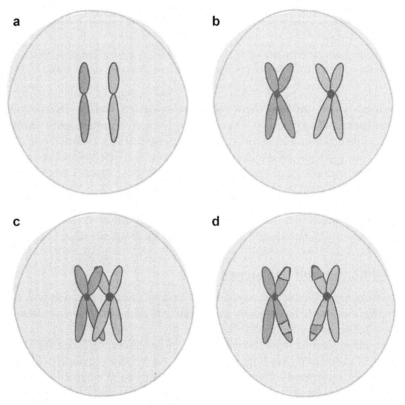

FIGURE 1.14 Crossing over and recombination during meiosis.

(A) Chromosomes come in pairs called homologous chromosomes. One chromosome in the pair is obtained from the mother (maternal), and the second from the father (paternal). (B) One of the first steps in meiosis is the replication of genetic material. (C) The homologous chromosomes align in pairs at the cell equator, and the arms of the chromosomes may cross over. (D) Crossing over results in the exchange of genetic material between the homologous chromosomes.

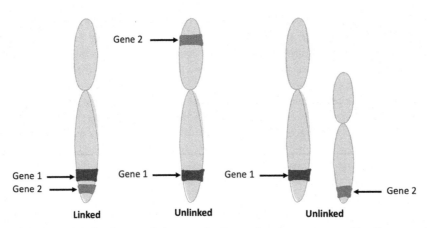

FIGURE 1.15 **Linked genes lie in close proximity to each other on the chromosome, making them more likely to be inherited together.**

Genes farther away from each other on the same chromosome are more likely to be separated during recombination and may not be linked. Genes on separate chromosomes are never linked.

Following recombination, the homologous pairs are separated, and the chromosomes are sorted randomly into daughter cells Fig. 1.16. This random division of homologous chromosomes into different daughter cells underlies the law of independent assortment, in which the alleles of different genes are sorted into gametes independently of one another. Recombination and independent assortment act during meiosis to produce nonidentical gametes, which causes offspring to have combinations of alleles different from their parents.

Meiosis is like preparing a deck of cards for a game. Recombination shuffles the combination of genes inherited together much like shuffling changes the combination of cards in a deck. The chromosomes are randomly sorted into daughter cells as cards are randomly dealt to players.

Linkage disequilibrium and haplotypes

From the Hardy-Weinberg principle one would expect that, in a large population with random mating and no selection, variants should be randomly associated with each other. However, this is not the case when variants are in linkage disequilibrium (LD).

LD is the nonrandom correlation between variants in proximity, such that the variants co-occur more frequently than expected by random assortment.[24] As we saw previously (Fig. 1.13), variants or SNPs that lie close together on a chromosome are more likely to be inherited together. A haplotype is the combination of alleles for different polymorphisms that reside near each other on a chromosome and tend to be inherited together Fig. 1.17. Recombination between variants within a haplotype is rare.

For any given stretch of chromosomal DNA, an individual will have two haplotypes (corresponding to the two chromosomes in a homologous pair). However, at the population level, there may be several different haplotypes for a given stretch of chromosomal DNA. Fig. 1.18 demonstrates this fact (obtained from Conrad et al. 2006).[25] Fig. 1.18 shows haplotype structures in different populations for a 330 kb genomic region. Each box represents a single population and haplotypes are plotted as thin

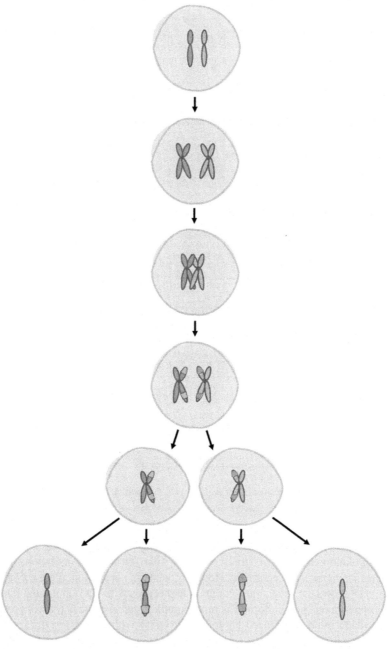

FIGURE 1.16 Meiosis is a specialized cell division that is important for sexual reproduction.

Recombination and independent assortment occur during meiosis and result in the daughter cells being genetically unique to each other and the parent cells.

	SNP	SNP	SNP	SNP	Haplotypes
Individual 1	A T G C **G** T T G A C G T	A G G C A **T** G T	C A C C C T G T	**T** A A G A C G T A	G A T T
Individual 2	G G C A **T** A T G C G T T	G A C G T A **A** G T	C A C C C T G T	**T** A A T T G A C G	T G A T
Individual 3	A G T C **T** T G C T A C A	A T G G C **C** T T	C C C C A T C C	**T** A C C T C T G C	T A C T

FIGURE 1.17 Haplotypes are composed of a unique combination of variants that reside near each other on a chromosome.

Haplotypes are a product of linkage disequilibrium (LD).

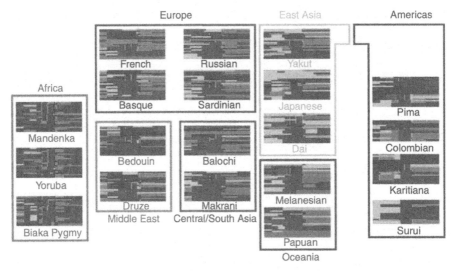

FIGURE 1.18

Haplotype diversity across different populations.

From Conrad DF, Jakobsson M, Coop G, Wen X, Wall JD, Rosenberg NA, Pritchard JK. A worldwide survey of haplotype variation and linkage disequilibrium in the human genome. Nature Genetics. *2006;38(11):1251–1260.*

horizontal lines. Different haplotypes are plotted with a different color, using the same haplotype coloring for all populations. Haplotype structures are similar for nearby populations, particularly for populations that reside on the same continent. However, within a population, there are several different haplotypes (Fig. 1.18 shows the seven most common haplotypes within a population, but there may be many more haplotypes in one population). The most complex and variable haplotypes reside in Africa, while there is a steady trend of reduced haplotype diversity with increasing distance from Africa. Additionally, LD in Africa (particularly sub-Saharan Africa) is reduced compared to that in other populations. This pattern of reduced haplotype diversity and increased LD outside of Africa stems from the migration of early humans outside of Africa.

Recombination breaks up haplotypes, causing ancestral haplotypes to become shorter with each successive generation. However, some regions of the human genome are more likely to undergo recombination than others. Regions of the genome where there is little evidence of genetic recombination, and which contain only a small number of discrete haplotypes, are called haplotype blocks.[26] LD within haplotype blocks is high. Haplotype blocks typically lie between regions of the genome with a high rate of recombination (termed recombination hotspots).[27] It is thus possible to characterize individuals by a few common haplotypes, which are parts of ancestral haplotypes that are conserved in the population.

Haplotype blocks can be leveraged in genetic studies to genotype individuals. When the recombination rate is low, the DNA sequence of the haplotype block will largely be conserved and any variation within haplotype blocks will largely be caused by mutation. The mutations will have been introduced in a population on a specific haplotype background. Instead of testing all SNPs, it suffices to test only those SNPs or combination SNPs that uniquely tag one of the common haplotypes. For example, consider Fig. 1.19. To characterize the haplotypic variation in haplotype block 2, only six of the 13 SNPs within the block need to be genotyped. The untyped SNPs can then be inferred from a reference panel (this process of inferring untyped SNPs is called imputation). Genome-wide association studies (GWASs) also leverage LD between genotyped tagging SNPs to identify causative SNPs.

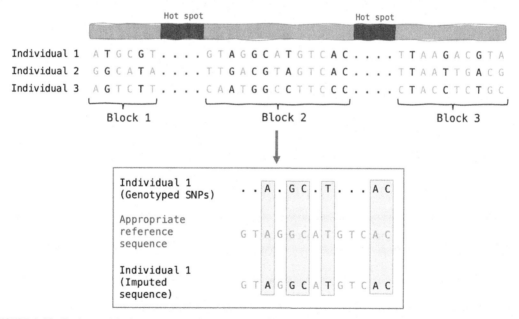

FIGURE 1.19 Haplotype blocks are conserved sequences of chromosomal DNA that do not undergo recombination as they are passed onto subsequent generations.

Haplotype blocks typically lie between recombination hot spots. Haplotype blocks can be characterized by genotyping only a few SNPs within the block. The untyped SNPs within the haplotype block can be inferred from an appropriate reference panel.

Measuring linkage disequilibrium

There are two measurements that can be used to quantify LD between two SNPs[28]: D' and r^2..

For two SNPs in complete LD (i.e., will never be separated by recombination), $D' = 1$ and $r^2 = 1$. Under complete LD, there will be at most three or four possible haplotypes in the population. The degree of LD between two SNPs can be calculated from the allele frequencies in the population.

Alternatively, LD heat maps can be used to visually represent the extent of LD among several SNPs in a region. In LD heat maps, blocks colored in red indicate regions in high LD. LD heat maps can also help elucidate the boundaries between haplotype blocks within a genomic region. An example of an LD heat map is provided in Fig. 1.20.[29]

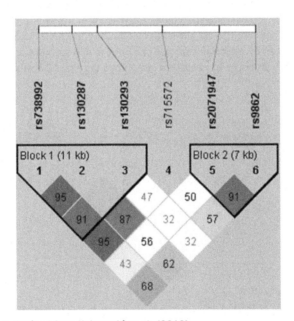

FIGURE 1.20 LD heat map obtained from Kobayashi et al. (2013).

Red blocks indicate regions in high LD. D' values that correspond to SNP pairs are expressed as percentages and are shown within the respective squares. Higher D' values are indicated with a bright red color. Two haplotype blocks can be identified in the diagram (surrounded by a black border) and a recombination hot spot with low levels of LD lies between the two haplotype blocks. SNPs rs738992, rs130287, and rs130293 are in LD in block 1. SNPs rs2071947 and rs9862 are in LD in block 2. For interpretation of the references to color in this figure legend, please refer online version of this title.

From Kobayashi N, Hanaoka M, Droma Y, Ito M, Katsuyama Y, Kubo K, Ota M. Polymorphisms of the tissue inhibitor of metalloproteinase 3 gene are associated with resistance to high-altitude pulmonary edema (HAPE) in a Japanese population: a case control study using polymorphic microsatellite markers. PloS One. 2013;8(8):e71993. https://doi.org/10.1371/journal. pone.0071993.

Factors influencing linkage disequilibrium

Several factors contribute to the extent of LD in the genome[30]:

Natural selection

Natural selection can produce LD. For example, a combination of alleles at two or more loci that contribute to the reproductive success of an individual are likely to be passed onto subsequent generations, thereby creating LD between these alleles. Natural selection may also affect the extent of LD. For example, if a variant within a haplotype block is favored by selection, the entire haplotype will be swept to high frequency or even fixation within the population.

Genetic recombination

As mentioned previously, recombination breaks up haplotype blocks. With successive generations and additional recombination events, the haplotype blocks become shorter and LD among alleles within the haplotype block decreases. The disruption of LD by recombination is known as LD decay.

Genetic drift

Genetic drift is a change in the allele frequencies of a population that occurs by chance and not by natural selection. Some alleles may be lost because of genetic drift, thereby causing other alleles to increase in frequency or become fixed in the population. Changes in allele frequencies are more pronounced in smaller populations; thus smaller populations will be more susceptible to genetic drift. Like natural selection, genetic drift will increase LD.

Population growth

Rapid population growth will decrease LD by reducing the effects of genetic drift.

Consanguinity

Consanguinity creates LD. Due to common ancestry, related individuals will have similar haplotype structures. High levels of inbreeding also increase the levels of homozygosity within a population. Homozygosity reduces the effectiveness of recombination—recombination creates new combinations of alleles, but new combinations cannot be created when alleles are identical. Subsequently, consanguinity and inbreeding preserve LD.

Admixture

Admixture occurs when individuals from two or more previously isolated populations interbreed. The previously isolated populations are referred to as the ancestral populations and the newly formed population is referred to as admixed. Admixed individuals are unique as their haplotypes have two levels of LD—ancestral-induced LD and admixture-induced LD. Ancestral-induced LD refers to the LD between polymorphisms within a conserved haplotype block, as described previously. However, admixture-induced LD is LD between loci (even unlinked loci) that is produced by admixture.

Fig. 1.21 demonstrates the difference between haplotypes produced by ancestral-induced LD and admixture-induced LD. In Fig. 1.21A, the ancestral haplotype is passed on from one generation to the next. The polymorphisms within the ancestral haplotype block are in LD. After successive recombination events, the haplotype shortens over time, and LD among the polymorphisms within the haplotype block decays. Fig. 1.21B is a basic diagram demonstrating admixture between two

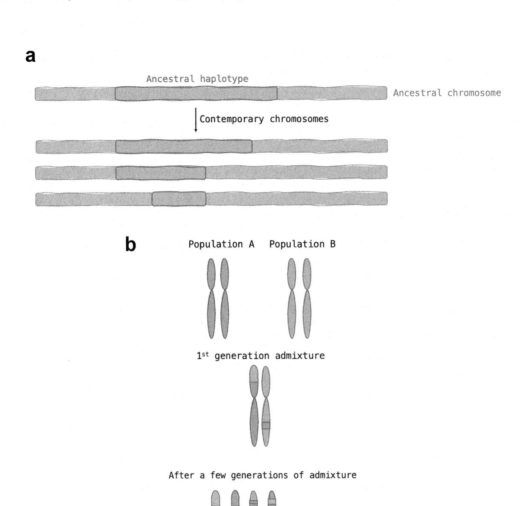

FIGURE 1.21 Admixed genomes have two different levels of linkage disequilibrium.

(A) Ancestral-induced LD is linkage disequilibrium among polymorphisms that lie within the same ancestral haplotype. With additional recombination events, the ancestral haplotype shortens. (B) Admixture between two populations produces linkage disequilibrium among polymorphisms within the admixture blocks. These polymorphisms may or may not have been linked or reside on the same haplotype prior to the admixture event.

individuals from different populations. After successive admixture events over several generations, the chromosomes of admixed individuals are a mosaic of chromosomal tracts originating from the two populations. Polymorphisms in the admixed-induced blocks are inherited together, regardless of if the loci are linked or unlinked or if they reside within the same ancestral haplotype.

Admixture-induced LD is leveraged to identify causative disease variants in admixture mapping. Admixture mapping is a method used to identify genetic variants associated with traits and/or diseases in ethnic groups whose genomes resulted from a recent mixture of two or more geographically distinct ancestral populations.

Identity by descent (IBD) and identity by state (IBS)

As discussed previously, individuals of the same ancestry will have tracts of their chromosomal DNA that are similar, since they are derived from the same ancestral haplotype block. When the chromosomal segment has been inherited from a common ancestor, then it is said to be identical by descent (IBD).[31] IBD is distinct from identity by state (IBS), in which a portion of two individual's genomes may appear identical by chance, and not necessarily due to recent shared co-inheritance.

IBS can be established by DNA sequencing. However, IBD can only be determined using a combination of DNA sequencing and inspection of the family pedigree (an example of a pedigree demonstrating IBD is shown in Fig. 1.22).

The more closely related two individuals are, the larger the percentage of their genome displays IBD since they share a common ancestor more recently than two randomly sampled individuals from the general population (Fig. 1.23). First-degree relatives (e.g., parents and children, full siblings) share 50% of their genome by descent. Second-degree relatives (e.g., grandparents and grandchildren, aunts/uncles and nephews/nieces, and half-siblings who share just one parent) share 25% of their genome by descent, on average. Third-degree relatives (e.g., first cousins) share 1/8th of their genome by descent, on average.

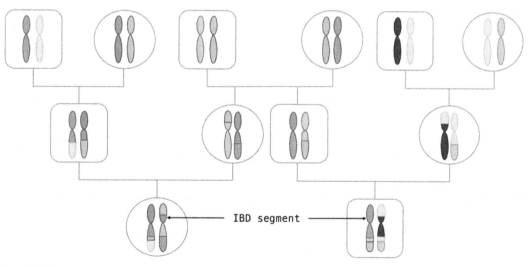

FIGURE 1.22

An individual's chromosomes resemble a mosaic of DNA sequences inherited from the previous generations. Individuals who share a common ancestor will have regions of the genome that are identical by descent.

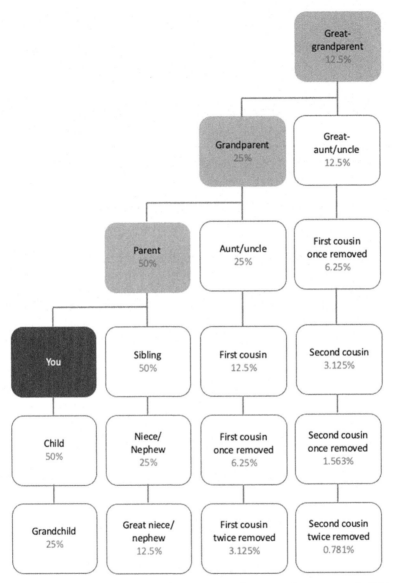

FIGURE 1.23 The percentage of genomic DNA identical by descent increases the closer two individuals are related to each other.

With successive generations and additional recombination events, the proportion of genomic DNA identical by descent is reduced.

As populations become more admixed over time, lengths of IBD segments will degrade due to recombination. Individuals that are closely related share longer haplotype blocks due to there being less recombination-induced LD decay.

Offspring from consanguineous unions typically have long runs of homozygosity in their genomes, in which the two homologous chromosomes of an individual are identical over long distances. Populations that have high rates of consanguineous unions will thus also have large fractions of their genome residing in long runs of homozygosity.

Applications of IBD in genetic studies

It is possible to infer events of a population's demographic history (such as population bottlenecks and founder effects) from IBD segments. The length of a shared IBD segment serves as a proxy for the age of the most recent common ancestor (MRCA) of that genomic region. For example, a long shared IBD segment reflects a more recent common ancestor. IBD segments can also be used to create population-specific recombination maps, which record the position of recombination hot spots and haplotype blocks in the genome.

Ancestral contributions, mtDNA, and Y-chromosome haplogroups
What is ancestry?

The term ancestry means different things to different people. For this reason, it is not easy to define. However, in the broad sense of the term, ancestry and in turn, ancestors signify relationships or connections to people or a population. Ancestors are therefore individuals you share a biological connection with, while ancestry is information/characteristics of these ancestors.

In the field of population genetics, the term ancestry is routinely used. There are, however, different forms of ancestry such as genealogical ancestry and genetic ancestry (or genetic similarity). These terms are distinct:

Genealogical ancestry

Genealogical ancestry is the ancestry as determined by inferring relationships to individuals. Therefore to estimate genealogical ancestry, there are three things that are required, namely, information regarding your ancestors, how they are all related (i.e., a detailed pedigree), and sufficient information to assign these ancestors to a particular geographical region or culture.

Genetic ancestry

Genetic ancestry makes use of genealogical ancestry in the sense that relationships between individuals are inferred, but related to doing this through genetic information, i.e., determining the paths by which your genome has been inherited from these ancestors.

Genetic similarity

Genetic similarity is what we routinely refer to as ancestry in population genetics and is based on the similarity between an individual and a set of reference/ancestral populations. Genetic similarity is then written as the percentage similarity in the form of admixture proportions.

How is genetic ancestry investigated?

In the field of genetics, the concept of the MRCA is described as the individual from which all other individuals are descended. For example, the MRCA of *Homo sapiens* (modern human) is *Homo erectus*.

Genetic variation (more specifically, haplotype variation) can be used to find this MRCA and to cluster individuals into groups with similar haplotypes, called haplogroups. In this way, if two individuals share a haplogroup, it would mean that they share a common ancestor.

Although autosomes can be used to establish haplogroups, it is more common to investigate this in the Y-chromosome and the mitochondrial (mtDNA) genome since they largely do not undergo recombination. Individuals that share common Y-chromosome and mtDNA haplotypes are said to belong to the respective haplogroup. The tracing of the development of new polymorphisms and therefore new haplogroups can be done, and even traced back to the MRCA of modern humans. For mtDNA, the MRCA can be found in southern Africa Fig. 1.24 and the oldest extant mtDNA haplogroup can be traced back to the KhoeSan.

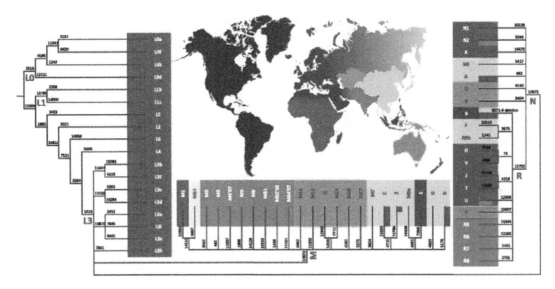

FIGURE 1.24 mtDNA haplogroup tree and distribution map.

The oldest extant mtDNA haplogroup is L0, which originated in southern Africa.

From Kivisild T. Maternal ancestry and population history from whole mitochondrial genomes. Investig Genet. *2015;6:3. https://doi.org/10.1186/s13323-015-0022-2.*

How is genetic similarity investigated?

Genetic similarity is often investigated through various statistical algorithms. These routinely include the estimation of global ancestry, local ancestry, or simply, variation clustering methods, e.g., principal component analysis (PCA). We will now discuss each of these concepts.

Global ancestry is the percentage of genome-wide similarity compared to other distinct ancestral populations. For example, 80% European and 20% African. Software tools such as ADMIXTURE[32,33] and STRUCTURE,[34] among others, are able to quantify global ancestry proportions with a mid-to-high degree of accuracy. A diagrammatical example of global ancestry is depicted below Fig. 1.25.

In contract, local ancestry is a bit more specific. Local ancestry estimations provide a specific ancestral population to which a particular chromosomal segment is most similar. In this way, each segment of a chromosome is assigned an ancestry based on similarity. Software tools such as RFMix,[35] ELAI,[36] and FLARE,[37] among others, are able to estimate local ancestry with a mid-to-high degree of accuracy. A diagrammatical example of local ancestry is depicted below Fig. 1.26.

Another method to quantify genetic similarity is in the form of a PCA. A PCA's main goal is to simplify the variability in large-scale datasets using dimensionality reduction methods. A diagrammatical example of a PCA is depicted below Fig. 1.27.

While the majority of the tests above made use of genotyping array data in the past, there is a significant movement toward making use of whole genome sequencing (WGS) data and even low coverage WGS data. Taking this a step further, we are also able to only genotype a set of SNPs that are statistically representative of global ancestry proportions. The reason for this is that SNPs exhibit substantially different frequencies between populations and are therefore distinct. These SNPs are called ancestry informative markers (AIMs). For example, the global ancestry proportions of a population using a set of 120 AIMs were similar to those estimated from a genotyping array including ~600,000 SNPs.[38] The more genetically complex a population, the more difficult it is to identify these AIMs, and generally, these populations will require more AIMs than more homogenous populations.

FIGURE 1.25 ADMIXTURE plot showing the genome-wide ancestral proportions of South African admixed individuals.

Ancestral proportions of each individual are plotted vertically. Ancestral populations are shown on the right.

From Swart Y, Uren C, van Helden PD, Hoal EG, Möller M. Local Ancestry Adjusted Allelic Association Analysis Robustly Captures Tuberculosis Susceptibility Loci. Front Genet. 2021;12:716558. https://doi.org/10.3389/fgene.2021.716558.

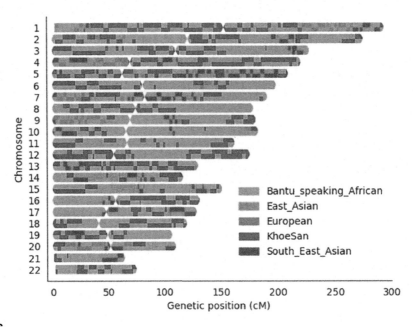

FIGURE 1.26

Karyogram showing the local ancestry proportions across all chromosomes for one admixed South African individual.

From Swart Y, Uren C, van Helden PD, Hoal EG, Möller M. Local ancestry adjusted allelic association analysis robustly captures tuberculosis susceptibility loci. Front Genet. 2021;12:716558. https://doi.org/10.3389/fgene.2021.716558.

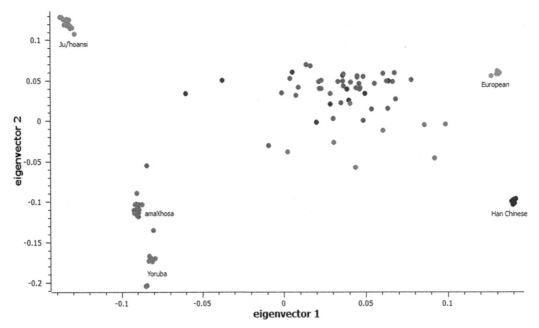

FIGURE 1.27 PC-plot showing clusters of individuals from similar population groups.

Samples that are more genetically similar based on specific SNPs genotyped will cluster closely together, such as African populations (amaXhosa and Yoruba).

References

1. Crow JF. Eighty years ago: the beginnings of population genetics. *Genetics.* 1988;119(3):473−476.
2. Hardy GH. Mendelian proportions in a mixed population. *Science (New York, NY).* 1908;28(706):49−50.
3. Crow JF. Hardy, Weinberg and language impediments. *Genetics.* 1999;152(3):821−825.
4. Edwards AWF. G. H. Hardy (1908) and Hardy-Weinberg equilibrium. *Genetics.* 2008;179(3):1143−1150. https://doi.org/10.1534/genetics.104.92940.
5. Wigginton JE, Cutler DJ, Abecasis GR. A note on exact tests of Hardy-Weinberg equilibrium. *Am J Hum Genet.* 2005;76(5):887−893.
6. Smocovitis VB. Unifying biology: the evolutionary synthesis and evolutionary biology. *J Hist Biol.* 1992; 25(1):1−65.
7. Epstein E, Guttman R. Mate selection in man: evidence, theory, and outcome. *Soc Biol.* 1984;31(3−4): 243−278.
8. Karlin S. Equilibrium behaviour of population genetic models with non-random mating. *J Appl Probab.* 1968;5(2):231−313. https://doi.org/10.2307/3212254.
9. Karlin S. Equilibrium behaviour of population genetic models with non-random mating: Part II: pedigrees, homozygosity and stochastic models. *J Appl Probab.* 1968;5(3):487−566. https://doi.org/10.2307/3211920.
10. Khlat M, Khoury M. Inbreeding and diseases: demographic, genetic, and epidemiologic perspectives. *Epidemiol Rev.* 1991;13:28−41.
11. Thiessen D, Gregg B. Human assortative mating and genetic equilibrium: an evolutionary perspective. *Ethol Sociobiol.* 1980;1(2):111−140. https://doi.org/10.1016/0162-3095(80)90003-5.
12. Ségurel L, Wyman MJ, Przeworski M. Determinants of mutation rate variation in the human germline. *Annu Rev Genom Hum Genet.* 2014;15:47−70. https://doi.org/10.1146/annurev-genom-031714-125740.
13. Serjeant GR. One hundred years of sickle cell disease. *Br J Haematol.* 2010;151(5):425−429. https://doi.org/10.1111/j.1365-2141.2010.08419.x.
14. Slatkin M. Gene flow and the geographic structure of natural populations. *Science (New York, NY).* 1987; 236(4803):787−792.
15. Masel J. Genetic drift. *Curr Biol: CB.* 2011;21(20):R837−R838. https://doi.org/10.1016/j.cub.2011.08.007.
16. Henn BM, Cavalli-Sforza LL, Feldman MW. The great human expansion. *Proc Natl Acad Sci USA.* 2012; 109(44):17758−17764. https://doi.org/10.1073/pnas.1212380109.
17. Zeegers MPA, van Poppel F, Vlietinck R, Spruijt L, Ostrer H. Founder mutations among the Dutch. *Eur J Hum Genet : EJHG (Eur J Hum Genet).* 2004;12(7):591−600.
18. Nevo E. Genetic variation in natural populations: patterns and theory. *Theor Popul Biol.* 1978;13(1): 121−177.
19. Anguita-Ruiz A, Aguilera CM, Gil Á. Genetics of lactose intolerance: an updated review and online interactive world maps of phenotype and genotype frequencies. *Nutrients.* 2020;12(9). https://doi.org/10.3390/nu12092689.
20. Lohmueller KE, Albrechtsen A, Li Y, et al. Natural selection affects multiple aspects of genetic variation at putatively neutral sites across the human genome. *PLoS Genet.* 2011;7(10). https://doi.org/10.1371/journal.pgen.1002326. e1002326.
21. Hedrick PW. Balancing selection. *Curr Biol: CB.* 2007;17(7):R230−R231.
22. Sanjak JS, Sidorenko J, Robinson MR, Thornton KR, Visscher PM. Evidence of directional and stabilizing selection in contemporary humans. *Proc Natl Acad Sci USA.* 2018;115(1):151−156. https://doi.org/10.1073/pnas.1707227114.
23. Griswold MD, Hunt PA, Maloy S, Hughes K. *San Diego: Academic Press.* 2013:338−341. https://doi.org/10.1016/B978-0-12-374984-0.00916-5.
24. Single RM, Thomson G, Kliman RM. *Oxford: Academic Press.* 2016:400−404. https://doi.org/10.1016/B978-0-12-800049-6.00030-5.

25. Conrad DF, Jakobsson M, Coop G, et al. A worldwide survey of haplotype variation and linkage disequilibrium in the human genome. *Nat Genet*. 2006;38(11):1251−1260.
26. Wall JD, Pritchard JK. Haplotype blocks and linkage disequilibrium in the human genome. *Nat Rev Genet*. 2003;4(8):587−597.
27. Liang K-H, Liang K-H. In: *Woodhead Publishing Series in Biomedicine*. Woodhead Publishing; 2013:9−48. https://doi.org/10.1533/9781908818232.9.
28. Calabrese B, Ranganathan S, Gribskov M, Nakai K, Schönbach C. *Oxford: Academic Press*. 2019:763−765. https://doi.org/10.1016/B978-0-12-809633-8.20234-3.
29. Kobayashi N, Hanaoka M, Droma Y, et al. Polymorphisms of the tissue inhibitor of metalloproteinase 3 gene are associated with resistance to high-altitude pulmonary edema (HAPE) in a Japanese population: a case control study using polymorphic microsatellite markers. *PLoS One*. 2013;8(8):e71993. https://doi.org/10.1371/journal.pone.0071993.
30. Slatkin M. Linkage disequilibrium–understanding the evolutionary past and mapping the medical future. *Nat Rev Genet*. 2008;9(6):477−485. https://doi.org/10.1038/nrg2361.
31. Thompson EA. Identity by descent: variation in meiosis, across genomes, and in populations. *Genetics*. 2013;194(2):301−326. https://doi.org/10.1534/genetics.112.148825.
32. Alexander DH, Lange K. Enhancements to the ADMIXTURE algorithm for individual ancestry estimation. *BMC Bioinf*. 2011;12:246. https://doi.org/10.1186/1471-2105-12-246.
33. Alexander DH, Novembre J, Lange K. Fast model-based estimation of ancestry in unrelated individuals. *Genome Res*. 2009;19(9):1655−1664. https://doi.org/10.1101/gr.094052.109.
34. Pritchard JK, Stephens M, Donnelly P. Inference of population structure using multilocus genotype data. *Genetics*. 2000;155(2):945−959.
35. Maples BK, Gravel S, Kenny EE, Bustamante CD. RFMix: a discriminative modeling approach for rapid and robust local-ancestry inference. *Am J Hum Genet*. 2013;93(2):278−288. https://doi.org/10.1016/j.ajhg.2013.06.020.
36. Guan Y. Detecting structure of haplotypes and local ancestry. *Genetics*. 2014;196(3):625−642. https://doi.org/10.1534/genetics.113.160697.
37. Browning SR, Waples RK, Browning BL. Fast, accurate local ancestry inference with FLARE. *Am J Hum Genet*. 2023;110(2):326−335. https://doi.org/10.1016/j.ajhg.2022.12.010.
38. Daya M, van der Merwe L, Galal U, et al. A panel of ancestry informative markers for the complex five-way admixed South African coloured population. *PLoS One*. 2013;8(12):e82224. https://doi.org/10.1371/journal.pone.0082224.

Why focus on population genomics in developing countries?

2

Anwani Siwada and Marlo Möller

South African Medical Research Council Centre for Tuberculosis Research, Division of Molecular Biology and Human Genetics, Faculty of Medicine and Health Sciences, Stellenbosch University, Cape Town, South Africa

Introduction

To gain a deeper insight of genetic variability at the individual, population, and continental level, population genomics broadens the scope of genomics research beyond reference data. Population genomics can assist with our understanding of how evolutionary processes have shaped variation across genomes and the global population.[1] According to statistics from the United Nations Conference on Trade and Development, developing economies have been at the center of the world's population growth over the last 25 years, with the proportion of humans who live in these regions increasing from 66% in 1950 to 83% in 2022 Fig. 2.1.[2] It is forecast that this percentage will increase to 86% in 2050. In contrast to this increase, most human genomics research has focused on European and Asian cohorts, largely from developed economies.[3] This is concerning, as the genetic data currently available are clearly not representative of the world's population. These datasets may also not be relevant to understudied groups and could even be unhelpful when determining genetic risk profiles for diseases in these settings, as a clear understanding of local population genomics is needed.[4] At the same time, developing economies, especially those in Africa, bear the brunt of socioeconomic inequalities, poor living conditions, and disease.[2] Primary healthcare incorporating genomics could assist developing economies in not only diagnosing and treating diseases but could also help to prevent diseases, resulting in cost savings.[5] However, to make precision and preventative medicine a reality for all, investigations of population genomics in diverse populations from across the world are crucial.

Research findings based on genomics have an impact on human lives.[6] This can be via medical developments, how we interact with the natural world, how we feel about the present and future of our species, and how we perceive ourselves and others.[6] The manner in which countries and people are classified therefore emphasizes a controversial problem that cannot be readily solved, as it is a topic that can be approached from many viewpoints.[7,8] On the one hand, scientists require classifications to organize and make sense of diversity, but on the other hand, such categorizations may create a false notion of superiority or hierarchy.[7] This was indeed the outcome of several terms that have been used to classify countries based on various metrics. These terms include but are not limited to developed

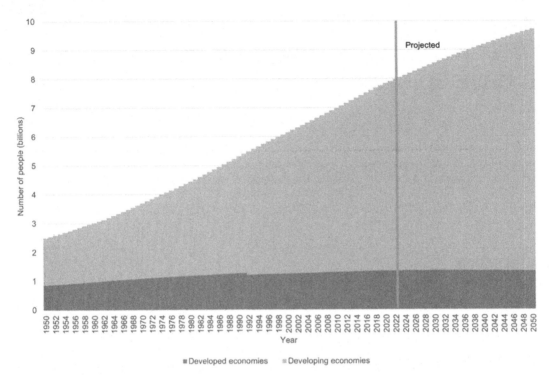

FIGURE 2.1 Developing economies are at the center of the world's population growth, as indicated by the distribution of the global population.

Data were retrieved on November 30, 2022 from UNCTADstat based on UN DESA population division, world population prospects 2022 (https://unctadstat.unctad.org).

versus developing countries or economies, high-income countries (HICs) against low- or middle-income countries (LMIC), Global North opposed to Global South, rich versus poor nations as well as resource-rich compared to resource-limited settings. A recent example of how both genomics and terminology use may have unintended consequences was the discovery of the Omicron coronavirus-19 variant of severe acute respiratory syndrome coronavirus 2 (SARS-CoV-2) by South African scientists.[9] The international response was to implement travel bans against many Southern African countries, even though it was subsequently shown that the variant did not originate in South Africa and was likely already present elsewhere by the time it was first reported.[10,11] South Africa is described as a developing country or LMIC, but it is clearly not poor in science or technology, as it was able to detect this new virus variant, even though this is an impression that might be created when this terminology is used.[8]

In this chapter, we will use the term that was used by the cited reference, but it is important to keep in mind that these classifications are based on historical, financial, or developmental metrics and may be problematic.

The importance of population genomics in developing countries

Population genomics has been used to study genetic ancestry as well as human health and migration patterns around the world. The field is a crucial area of study that should be prioritized in developing countries for the following six reasons.

Evolutionary biology

As we aim to understand the origins and history of our species, the study of human evolution has garnered a lot of attention. Developing countries often serve as hotspots for evolutionary studies due to their unique demographic histories, environmental conditions, and selection pressures. Investigating population genomics in these regions can enhance our understanding of human evolution,[12] migration patterns,[13–15] and the impact of natural selection. These insights have implications not only for addressing questions related to adaptation, genetic resilience, and disease evolution but also for understanding our shared human history.

Although it is now generally accepted that humans evolved in Africa, several opposing conceptual models representing our origins have been proposed. The "tree of life," one of biology's core organizing ideas, served as the main paradigm for human evolution for a long time. The genetic evidence was consistent with this scenario of human population separation from a single African ancestral population, but it was challenging to reconcile with the evidence of *Homo sapiens* fossils and archaeological sites spread across the continent.[12] Fossils and archaeological records indicate the presence of anatomically modern humans across Africa between 300,000 and 100,000 years ago. Several key fossils, such as those found at Jebel Irhoud in Morocco, Herto in Ethiopia, and Klasies River in South Africa, demonstrate that anatomical features that originated in *Homo sapiens* were present throughout the continent during this period. Additionally, archaeological sites associated with *Homo sapiens*, particularly from the Middle Stone Age, are widely distributed across Africa, including the site with the oldest footprint identified.[16] The question of whether these populations represent the direct ancestors of contemporary humans or represent isolated local populations remained unanswered. Due to recent advances in population genetic tools, more complex modeling and inference using larger datasets have become possible.[17]

A recent study used these tools and shed light on the divergence of human populations on the African continent and challenged traditional models, suggesting a new framework—represented by a weakly structured stem—where stem populations separated, but continually exchanged genetic material Fig. 2.2.[12] This work would not have been possible without sequencing the most genetically diverse human genomes in the world—contemporary African DNA is key to understanding deep human history.

According to genetic and paleoanthropological evidence, the great demic (demographic and geographic) expansion started in Africa between 45,000 and 60,000 years ago and quickly led to human habitation of nearly all of the Earth's habitable regions, the so-called Out-of-Africa hypothesis.[18] Genomic evidence from modern humans suggests that this expansion was accompanied by a steady loss of genetic diversity, a phenomenon known as the "serial founder effect" (Fig. 2.3).[19] The genetics of human parasites, morphology, and languages are increasingly used to support the serial founder effect paradigm in addition to genomic data. The two fundamental characteristics of genetic variety in humans were created by this specific population history: genomes from the substructured

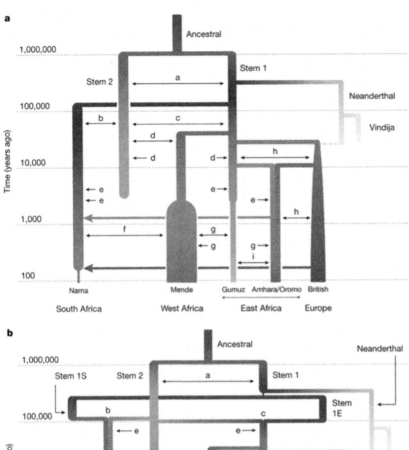

FIGURE 2.2

New "weakly structured stem" frameworks of human evolution with (a) continuous migration and (b) multiple mergers: Where stem populations separated, but continually exchanged genetic material.

From Ragsdale AP, Weaver TD, Atkinson EG, et al. A weakly structured stem for human origins in Africa. Nature. *2023;617(7962): 755–763. doi:10.1038/s41586-023-06055-y and licensed under CC BY 4.0 (https://creativecommons.org/licenses/by/4.0/).*

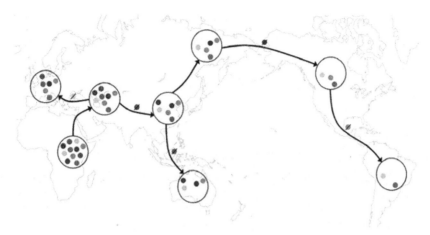

FIGURE 2.3

A schematic of the serial founder model in human evolution. Each colored dot represents genetic diversity. Each new population outside of Africa is a sample of the genetic diversity that existed in its founder population, as some diversity was lost during migration.

From Rosenberg NA, Kang JTL. Genetic diversity and societally important disparities. Genetics. 2015;201(1):1–12.
https://doi.org/10.1534/genetics.115.176750 by permission of Oxford University Press.

populations of Africa preserve an exceptionally high number of distinct genetic variations,[20] and genetic diversity has decreased dramatically in populations living outside Africa.

Clearly, genetic information from modern individuals still has a lot to teach us, especially in situations where ancient DNA, which is essential for shedding light on the fascinating history and providing answers to pressing questions, may not be present, as is sometimes the case in developing countries in Africa.

Genetic diversity

Developing countries often have higher levels of genetic diversity compared to developed nations. This diversity is due to various factors, including large and diverse populations, historical migration patterns, and unique evolutionary pressures. As mentioned in section Evolutionary biology, evolutionary studies rely on the Out-of-Africa hypothesis. *Homo sapiens* moved to different parts of the world and were exposed to different selection pressures, including differences in temperature and exposure to different diseases. This has contributed greatly to the great genetic differences developed within the populations in Africa as well as other evolutionary forces such as genetic drift. Even so, genomics research is biased toward European and Asian ancestry cohorts, and results from these investigations may not be applicable to individuals with more diverse ancestries. To prevent further disparities, and to improve knowledge about human history, health, and disease, it is therefore imperative to include diverse genetic datasets that have been generated using the newest technologies.

Following the initial major genomics project, The Human Genome Project (HGP), during which the first human reference genome was published in 2001,[21] there has been exponential growth in genomics studies producing great results. Until recently, the human genome reference was a widely accepted representation of the human genome sequence that researchers used as a benchmark when comparing it to DNA sequences produced for their studies.[21] However, it was derived from 11 individuals, with 70% of the DNA coming from one man, and biased toward European ancestry.[22] The reference genome has improved since its first release in 2001, including the release of a complete telomere-to-telomere version, which provided gapless assemblies for all chromosomes due to the technological ability to sequence regions that previous sequencing technologies could not process.[23,24] Even so, the underrepresentation of other populations in the reference genome had practical ramifications for those populations as it affects the ability to truly detect meaningful variation in an individual's genome sequence. Recently the initial genome was succeeded by a draft pangenome, a set of genomes from many individuals positioned to indicate where the sequences are matching or different.[25] This version of the human genome consists of 47 phased, diploid assemblies from genetically diverse individuals, added 119 million bases to the existing reference and was funded by the U.S. National Human Genome Research Institute at a cost of $30 million.[25]

Many countries, which today are mostly developing countries, were colonized by European countries (see Fig. 2.4) and this is not only significant in studies of history but also in genetic studies. This migration of people across the world resulted in many admixed groups, for example, many populations found in the southern part of Africa and in Brazil have ancestries from multiple continents.

FIGURE 2.4

A world map indicating the different countries colonized by different European countries.[26] The map was created using mapchart.net (https://www.mapchart.net/world.html).

From European Colonialism 1500 AD to 2000. (n.d.). Retrieved 2022, from https://commons.princeton.edu/mg/european-colonialism-1500-ad-to-2000/.

In South Africa up to five-way genetic admixture is present and includes Southeast Asian, European, Khoe-San, and Bantu-speaking ancestry.[27–29]

Southern Africa is also home to indigenous populations that form part of a larger group of geographically close and culturally related individuals known collectively as the "Khoe-San." The Khoe-San are reported to have the most divergent lineages of any other living population grouping and it is believed that they have largely remained isolated until ~2000 years ago, limiting exchanges with other populations and allowing for unique genetic variations to persist. In fact, studies in these groups have revealed thousands of novel variants that were not present in other populations.[20,30,31]

In recent years, there have been initiatives that focus on the development of genomics studies in developing countries and these ongoing initiatives have progressed and produced results that support the importance of genomics in developing countries (see Chapter 11). Initiatives taking place in Mexico, India, and Thailand[32] and the Genome for Greece (Go Greece), African Genome Variation Project (AGVP),[33] DNA do Brasil (DNABr),[34] and the Human Heredity and Health in Africa (H3Africa) Consortium[35] have been instrumental in collecting genomic data from diverse populations in developing countries.

The H3Africa initiative was put in place to facilitate research in genetic diseases that affect African populations. Great strides have been achieved so far by the initiative with regard to increasing the genomics knowledge of African populations. One of the most notable outcomes of the initiation is the development of a genotyping array, the H3Africa Array, which is an array that accounts for the genetic diversity and small haplotype blocks in African populations unlike other arrays developed based mostly on non-African populations.[36] A recent study that sequenced 180 individuals from 12 indigenous African populations identified millions of unreported variants, many with functional consequences.[37] The GoGreece initiative was initiated in 2010 with the aim to implement genomics in Greece. They aim to understand the Hellenic population genomes and have implemented the use of next-generation sequencing (NGS) technology in their various studies. Even so, there are definitely not enough genomes from developing countries sequenced (from modern or archaic samples) to represent genetic diversity, but there is a drive-by scientist to correct this.[38]

Studying population genomics in developing countries can provide valuable insights into the genetic variations that underlie diseases, traits, and drug responses, leading to more accurate and effective healthcare interventions globally.

Disease burden

The sequencing of the genomes of microorganisms, including SARS-CoV-2, *Mycobacterium tuberculosis*, and the influenza virus, has been the primary method for realizing the more immediate health benefits of investing in genomic sciences, both worldwide and in developing nations. Many developing countries face a significant burden of infectious diseases, but noncommunicable diseases such as diabetes, cardiovascular disorders, and certain types of cancer are also increasing. Population genomics research can aid in understanding the genetic basis of these diseases, identifying high-risk populations, and developing targeted prevention and treatment strategies.[39] It can also shed light on host-pathogen interactions and contribute to the development of vaccines and therapeutics.

One of the most popular study designs to investigate genetic disease associations is the case-control association study. Initially, these studies focused on single-candidate genes and compared allele frequencies between those with the disease ("cases") and those without the disease ("controls"). These

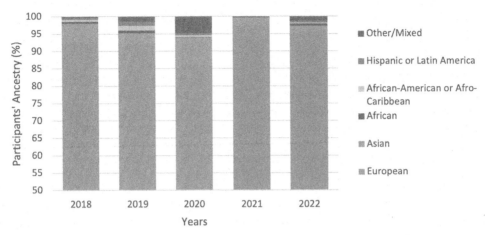

FIGURE 2.5

The ancestry of participants included in GWAS studies reported in the NHGRI-EBI catalog of human genome-wide association studies over recent years.

Information can be found on https://gwasdiversitymonitor.com/.

single-candidate gene studies were superseded by genome-wide association studies (GWAS), which predominantly focused on European ancestry samples (Fig. 2.5). Factoring the differences in allele frequencies between populations is crucial because variants are not equally common in all populations and progressing with this assumption may lead to overestimation or underestimation of disease risk in other populations.

It is reassuring to see that multi-ancestry GWAS are becoming more common in the literature.[40-42] One of these studies focused on elevated blood lipid levels, a heritable factor of cardiovascular disease, that differs between populations because of different diets and medications.[40] Not only is one-third of all deaths worldwide caused by cardiovascular diseases, but nearly 80% occur in developing nations.[32] Prior GWAS of elevated blood lipid levels contributed to new drug targets for cardiovascular disease and helped to elucidate the underlying biology of the phenotype. However, due to the uniform ancestry of participants in these initial studies, the genetic variants that influence this phenotype in other ancestries might have been missed. This is likely due to differences in linkage disequilibrium (LD) patterns, effect sizes, and allele frequencies. Instead, the authors studied approximately 1.65 million individuals, including 350,000 non-European participants, and made several key findings that could be extended to other complex phenotypes[40]:

The number of significant genetic regions associated with traits was similar within each ancestry group, and it was found to be linearly related to the sample size. African ancestry cohorts showed a slightly higher number of ancestry-specific variants compared to other groups.

Incorporating multiple ancestries in the analysis through trans-ancestry fine-mapping led to a reduction in the number of potential causal variants for a given trait. This reduction was observed in credible sets, which are sets of variants with a high probability of being causally linked to the trait. Trans-ancestry analysis allowed for more rapid identification of candidate causal variants compared to analyzing each ancestry separately.

A polygenic score, derived from many individuals with African and European ancestry, was correlated with observed lipid levels in individuals of mixed African ancestry as well as individuals of European ancestry. This suggests that including individuals with African ancestry in GWAS improves the predictive performance of polygenic scores. The improved prediction is thought to result from the finer mapping of genetic loci and better prioritization of trans-ancestry causal variants.

Finally, the study showed that the trans-ancestry polygenic score was informative across various population groups. This finding has important implications for other genetic studies and the use of polygenic scores in diverse populations, providing valuable information for future genetic discoveries and investigations into the utility of polygenic scores.

These findings have indeed already been shown to be applicable to other complex phenotypes. The International Tuberculosis (TB) Host Genetics Consortium (ITHGC)[43] performed a multi-ancestry meta-analysis GWAS that included 14,153 TB cases and 19,536 controls of African, Asian, and European ancestry from both developed and developing countries.[41] A substantial degree of heritability was shared between ancestries and one genome-wide significant association was identified in the human leukocyte antigen class II region. In an opposing view, another study involving TB and type 2 diabetes showed how ancestry-specific expression quantitative loci (eQTLs) might have been missed if the genetic ancestry of the study participants were not taken into consideration[44] and this was also shown in a TB GWAS.[45] Clearly, population genomics research can help to understand the genetic basis of diseases and identify populations at high risk of these diseases.

Precision medicine and pharmacogenomics

Precision medicine aims to provide tailored healthcare based on an individual's genetic makeup, lifestyle, and environmental factors. In developing countries, where resources may be limited, precision medicine can help avoid unnecessary treatments and reduce healthcare costs by targeting interventions to those who are most likely to benefit. This inclusion is crucial because genetic differences between populations can also influence disease susceptibility, response to medications, and treatment outcomes. Neglecting these populations could lead to health disparities and hinder the equitable implementation of precision medicine.

The fact that "common variants" are not common in all populations can be seen in some Mendelian disorders where the common mutations of the one disorder are frequent in the different populations (Table 2.1).[46] An example of this is sickle cell anemia, a form of the inherited blood disorder sickle cell disease (SDC), in African populations. Sickle cell anemia is a disease caused by a homozygous b^S mutation, which produces a mutated structural variant of adult hemoglobin known as sickle hemoglobin (HbS).[47] In most populations the b^S mutation occurs in low frequencies; however, studies have found that the allele frequency is specifically high in sub-Sahara African populations found in malaria-endemic areas. Positive selection has caused the allele frequency to be high because the allele in a heterozygous form has protective qualities against malaria. The GoGreece initiative was able to identify unique variants associated with amyotrophic lateral sclerosis (ALS) and pediatric celiac disease in Greek patients.[34]

The clinical utilization of genomic research results derived from studies that are not a representation of all global populations will result in diagnostic technology that is not as useful or could rather

Table 2.1 Comparison of mutations for monogenic disorders between European and African ancestry populations.

Disease	Gene	Common African mutation	Common European mutation
Cystic fibrosis	CFTR	3120+1G > A	ΔF508
Galactosemia	GALT	S135L (c.404C > T)	Q188R (c.563A > G)
Osteogenesis imperfecta type 3	FKBP10	c.831dupC	Diverse mutations
Glutaric aciduria	GCDH	A293T (c.877G > A)	R402W (c.1204C > T)

Adapted from, Krause A, Seymour H, Ramsay M. Common and founder mutations for monogenic traits in sub-saharan African populations. Annu Rev Genom Hum Genet. 2018;19:149–175. https://doi.org/10.1146/annurev-genom-083117-021256.

be harmful to individuals of other populations. It can also help to improve drug selection—by identifying genetic markers that influence drug metabolism and effectiveness, pharmacogenomics can help healthcare providers in developing countries choose the most appropriate medications and dosages for their patients. This can enhance treatment outcomes and minimize adverse reactions, especially in populations with unique genetic characteristics. Developing countries can benefit from these advancements by gaining access to more efficient drugs and therapies that are tailored to their populations, reducing the risk of adverse events and improving treatment outcomes. In developing countries, where limited resources are often allocated to treating advanced diseases, focusing on prevention and early detection can have a significant impact on public health. Incorporating precision medicine and pharmacogenomics in developing countries can stimulate research collaborations and capacity-building efforts. This can include establishing biobanks, genetic databases, and research networks, which contribute to the understanding of population-specific genetic variations and their impact on health and disease.

Recent examples of large-scale population genotyping projects in Mexico, India, Thailand, and Columbia include the Mexican National Genomic Medicine Institute (INMEGEN), the Indian Genome Variation database Consortium (IGVdb Consortium) and two initiatives in Thailand: the Thailand Single Nucleotide Polymorphism (SNP) Discovery Project and the Pharmacogenomics Project at the Thailand Center for Excellence in Life Sciences.[32,48] In Colombia, the study focused on two nearby populations with unique ancestry—one with predominantly European, Native American, and African genetic ancestry and another with mainly African ancestry. Interestingly, a number of pharmacogenomic variants had different effect allele frequencies within and between these two populations and this was associated with the genetic ancestry profiles of the two groups.[48] This was also found in a review of pharmacogenetic variants involved in TB treatment.[49] Although genomic approaches to precision medicine show enormous promise, their cost is often touted as too high for developing nations to use—although the question can also be asked if developing countries can afford to not use precision medicine.

Genomic research infrastructure

Investing in population genomics research in developing countries can contribute to the development of genomic research infrastructure and capacity building in these regions. It helps foster collaborations between local scientists and international research institutions, promoting knowledge transfer, technical expertise, and the establishment of sustainable scientific networks. This capacity building strengthens the scientific ecosystem, facilitates knowledge exchange, and ultimately benefits the local healthcare systems and population.

Genomics studies have advanced over recent years with the introduction of cutting-edge technology and study designs such as GWAS, NGS, and long-read sequencing technologies. With NGS technology such as whole genome sequencing (WGS), significantly more novel variants have been discovered and GWAS studies have been key to associate these variants with various complex phenotypes. The broad knowledge gained from genomic studies has translated to improved diagnostics and therapeutics of many diseases in clinical settings. Furthermore, this advancement has largely influenced the progression in the new field of precision medicine where disease-preventative measures, diagnosis, and treatments are tailored to an individual's genetic makeup, lifestyle, and environment.[50]

The establishment of local sequencing facilities on the African continent is another important step, as some African countries do not allow the export of DNA samples. In South Africa, we are fortunate to have access to several sequencing facilities, such as the South African Medical Research Council Genomics Platform,[51] as well as the Center for Epidemic Response and Innovation (CERI) Genomics Center at Stellenbosch University, to name but two. While significant progress has been made in recent years, it is important to note that the sequencing of genomes from developing countries using local infrastructure is an ongoing process, and there is still much work to be done. Access to generate data on diverse genomes in-country will advance our knowledge of human evolution and improve our analyses of genomic variation linked to complex health traits.

The impact of population genomics on developing countries

The importance of science and technology for human growth is becoming increasingly clear, and emerging nations are paying particular attention to the life sciences. Emerging economies and other developing nations are starting to construct infrastructures for domestic innovation and harness the value of their scientific research in an effort to break the cycle of reliance.[32] Countries that have implemented genomics have seen the economic impact of this field, including job creation and improved healthcare. In developing countries, this would have an impact on improving life, as there is a high disease burden and unemployment rates. Even so, the realization of advancements in genomics and precision medicine in developing countries is still fairly behind compared to developed countries, but examples such as GoGreece and DNABr provide evidence that population genomics can contribute to making genomic medicine in developing countries a reality.

Conclusions

By focusing on population genomics in developing countries, we can advance our understanding of human genetics, improve healthcare outcomes, reduce health disparities, and contribute to global scientific progress.

References

1. Luikart G, England PR, Tallmon D, Jordan S, Taberlet P. The power and promise of population genomics: from genotyping to genome typing. *Nat Rev Genet*. 2003;4(12):981−994. https://doi.org/10.1038/nrg1226.
2. Now 8 billion and counting: Where the world's population has grown most and why that matters. https://unctad.org/data-visualization/now-8-billion-and-counting-where-worlds-population-has-grown-most-and-why. Accessed 2022.
3. Sirugo G, Williams SM, Tishkoff SA. The missing diversity in human genetic studies. *Cell*. 2019;177(1): 26−31. https://doi.org/10.1016/j.cell.2019.02.048.
4. Martin AR, Teferra S, Möller M, Hoal EG, Daly MJ. The critical needs and challenges for genetic architecture studies in Africa. *Curr Opin Genet Dev*. 2018;53:113−120. https://doi.org/10.1016/j.gde.2018.08.005.
5. Developing genetics for developing countries. *Nat Genet*. 2007;39(11):1287. https://doi.org/10.1038/ng.2007.500.
6. Templeton AR. *Human Population Genetics/Genomics and Society*. Elsevier BV; 2019:437−473. https://doi.org/10.1016/b978-0-12-386025-5.00014-2.
7. Khan T, Abimbola S, Kyobutungi C, Pai M. How we classify countries and people—and why it matters. *BMJ Glob Health*. 2022;7.
8. Lencucha R, Neupane S. The use, misuse and overuse of the 'low-income and middle-income countries' category. *BMJ Glob Health*. 2022;7(6):e009067. https://doi.org/10.1136/bmjgh-2022-009067.
9. Viana R, Moyo S, Amoako DG, et al. Rapid epidemic expansion of the SARS-CoV-2 Omicron variant in southern Africa. *Nature*. 2022;603(7902):679−686. https://doi.org/10.1038/s41586-022-04411-y.
10. Mallapaty S. Where did Omicron come from? Three key theories. *Nature*. 2022;602(7895):26−28. https://doi.org/10.1038/d41586-022-00215-2.
11. Mallapaty S. Omicron-variant border bans ignore the evidence, say scientists. *Nature*. 2021;600(7888):199. https://doi.org/10.1038/d41586-021-03608-x.
12. Ragsdale AP, Weaver TD, Atkinson EG, et al. A weakly structured stem for human origins in Africa. *Nature*. 2023;617(7962):755−763. https://doi.org/10.1038/s41586-023-06055-y.
13. Bustamante CD, Henn BM. Shadows of early migrations. *Nature*. 2010;468(7327):1044−1045. https://doi.org/10.1038/4681044a.
14. Hunter P. The genetics of human migrations. *EMBO Rep*. 2014;15(10):1019−1022. https://doi.org/10.15252/embr.201439469.
15. Wolinsky H. Our history, our genes: population genetics lets researchers look back in time at human migrations. *EMBO Rep*. 2008;9(2):127−129. https://doi.org/10.1038/sj.embor.7401164.
16. Helm CW, Carr AS, Lockley MG, et al. Dating the Pleistocene hominin ichnosites on South Africa's Cape south coast. *Ichnos*. 2023:1−20. https://doi.org/10.1080/10420940.2023.2204231.
17. Scerri EML, Thomas MG, Manica A, et al. Did our species evolve in subdivided populations across Africa, and why does it matter? *Trends Ecol Evol*. 2018;33(8):582−594. https://doi.org/10.1016/j.tree.2018.05.005.
18. Henn BM, Cavalli-Sforza LL, Feldman MW. The great human expansion. *Proc Natl Acad Sci USA*. 2012; 109(44):17758−17764. https://doi.org/10.1073/pnas.1212380109.
19. Rosenberg NA, Kang JTL. Genetic diversity and societally important disparities. *Genetics*. 2015;201(1): 1−12. https://doi.org/10.1534/genetics.115.176750.
20. Schuster SC, Miller W, Ratan A, et al. Complete Khoisan and Bantu genomes from southern Africa. *Nature*. 2010;463(7283):943−947. https://doi.org/10.1038/nature08795.
21. Lander ES, Linton LM, Birren B, et al. Initial sequencing and analysis of the human genome. *Nature*. 2001; 409(6822):860−921. https://doi.org/10.1038/35057062.

22. Bien SA, Wojcik GL, Hodonsky CJ, et al. The future of genomic studies must be globally representative: perspectives from PAGE. *Annu Rev Genom Hum Genet.* 2019;20:181−200. https://doi.org/10.1146/annurev-genom-091416-035517.

23. Nurk S, Koren S, Rhie A, et al. The complete sequence of a human genome. *Science.* 2022;376(6588): 44−53. https://doi.org/10.1126/science.abj6987.

24. Rhie A, Nurk S, Cechova M, et al. The complete sequence of a human Y chromosome. *bioRxiv.* December 2022. https://doi.org/10.1101/2022.12.01.518724.

25. Liao WW, Asri M, Ebler J, et al. A draft human pangenome reference. *Nature.* 2023;617(7960):312−324. https://doi.org/10.1038/s41586-023-05896-x.

26. European Colonialism 1500 AD to 2000. https://commons.princeton.edu/mg/european-colonialism-1500-ad-to-2000/. Accessed 2022.

27. De Wit E, Delport W, Rugamika CE, et al. Genome-wide analysis of the structure of the South African coloured population in the Western Cape. *Hum Genet.* 2010;128(2):145−153. https://doi.org/10.1007/s00439-010-0836-1.

28. Petersen DC, Libiger O, Tindall EA, et al. Complex patterns of genomic admixture within southern Africa. *PLoS Genet.* 2013;9(3):e1003309. https://doi.org/10.1371/journal.pgen.1003309.

29. Choudhury A, Ramsay M, Hazelhurst S, et al. Whole-genome sequencing for an enhanced understanding of genetic variation among South Africans. *Nat Commun.* 2017;8(1). https://doi.org/10.1038/s41467-017-00663-9.

30. Schlebusch CM, Skoglund P, Sjödin P, et al. Genomic variation in seven Khoe-San groups reveals adaptation and complex African history. *Science.* 2012;338(6105):374−379. https://doi.org/10.1126/science.1227721.

31. Uren C, Kim M, Martin AR, et al. Fine-scale human population structure in Southern Africa reflects eco-geographic boundaries. *Genetics.* 2016;204(1):303−314. https://doi.org/10.1534/genetics.116.187369.

32. Séguin B, Hardy BJ, Singer PA, Daar AS. Genomic medicine and developing countries: creating a room of their own. *Nat Rev Genet.* 2008;9(6):487−493. https://doi.org/10.1038/nrg2379.

33. Gurdasani D, Carstensen T, Tekola-Ayele F, et al. The African genome variation project shapes medical genetics in Africa. *Nature.* 2015;517(7534):327−332. https://doi.org/10.1038/nature13997.

34. Patrinos GP, Pasparakis E, Koiliari E, et al. Roadmap for establishing large-scale genomic medicine initiatives in low- and middle-income countries. *Am J Hum Genet.* 2020;107(4):589−595. https://doi.org/10.1016/j.ajhg.2020.08.005.

35. Choudhury A, Aron S, Botigué LR, et al. High-depth African genomes inform human migration and health. *Nature.* 2020;586(7831):741−748. https://doi.org/10.1038/s41586-020-2859-7.

36. Mulder N, Abimiku A, Adebamowo SN, et al. H3Africa: current perspectives. *Pharmacogenomics Personalized Med.* 2018;11:59−66. https://doi.org/10.2147/PGPM.S141546.

37. Fan S, Spence JP, Feng Y, et al. Whole-genome sequencing reveals a complex African population demographic history and signatures of local adaptation. *Cell.* 2023;186(5):923−939.e14. https://doi.org/10.1016/j.cell.2023.01.042.

38. Wonkam A. Sequence three million genomes across Africa. *Nature.* 2021;590(7845):209−211. https://doi.org/10.1038/d41586-021-00313-7.

39. Karlsson EK, Kwiatkowski DP, Sabeti PC. Natural selection and infectious disease in human populations. *Nat Rev Genet.* 2014;15(6):379−393. https://doi.org/10.1038/nrg3734.

40. Graham SE, Clarke SL, Wu KHH, et al. The power of genetic diversity in genome-wide association studies of lipids. *Nature.* 2021;600(7890):675−679. https://doi.org/10.1038/s41586-021-04064-3.

41. Schurz H, Naranbhai V, Yates TA, et al. Multi-ancestry meta-analysis of host genetic susceptibility to tuberculosis identifies shared genetic architecture. *medRxiv.* August 2022. https://doi.org/10.1101/2022.08.26.22279009.

42. Mahajan A, Go MJ, Zhang W, et al. Genome-wide trans-ancestry meta-analysis provides insight into the genetic architecture of type 2 diabetes susceptibility. *Nat Genet*. 2014;46(3):234−244. https://doi.org/10.1038/ng.2897.

43. Naranbhai V. The role of host genetics (and genomics) in tuberculosis. *Microbiol Spectr*. 2016;4(5). https://doi.org/10.1128/microbiolspec.TBTB2-0011-2016.

44. Swart Y. *Cis-eQTL Mapping of TB-T2D Comorbidity Elucidates the Involvement of African Ancestry in TB Susceptibility*. 2022. https://doi.org/10.1101/2022.10.19.512814.

45. Swart Y, Uren C, van Helden PD, Hoal EG, Möller M. Local ancestry adjusted allelic association analysis robustly captures tuberculosis susceptibility loci. *Front Genet*. 2021;12. https://doi.org/10.3389/fgene.2021.716558.

46. Krause A, Seymour H, Ramsay M. Common and founder mutations for monogenic traits in sub-saharan African populations. *Annu Rev Genom Hum Genet*. 2018;19:149−175. https://doi.org/10.1146/annurev-genom-083117-021256.

47. Williams TN. Sickle cell disease in sub-saharan Africa. *Hematol Oncol Clin N Am*. 2016;30(2):343−358. https://doi.org/10.1016/j.hoc.2015.11.005.

48. Nagar SD, Moreno AM, Norris ET, et al. Population pharmacogenomics for precision public health in Colombia. *Front Genet*. 2019;10. https://doi.org/10.3389/fgene.2019.00241.

49. Oelofse C, Ndong Sima CAA, Möller M, Uren C. Pharmacogenetics as part of recommended precision medicine for tuberculosis treatment in African populations: could it be a reality? *Clin Transl Sci*. January 2023. https://doi.org/10.1111/cts.13520.

50. Zhang J, Späth SS, Marjani SL, Zhang W, Pan X. Characterization of cancer genomic heterogeneity by next-generation sequencing advances precision medicine in cancer treatment. *Precis Clin Med*. 2018;1(1):29−48. https://doi.org/10.1093/pcmedi/pby007.

51. Glanzmann B, Jooste T, Ghoor S, et al. Human whole genome sequencing in South Africa. *Sci Rep*. 2021;11(1). https://doi.org/10.1038/s41598-020-79794-x.

Contribution of large consortium studies to population genomics in the developing world: Examples from Africa and South Asia

Dhriti Sengupta[1], Analabha Basu[2], Michèle Ramsay[1] and Ananyo Choudhury[1]

[1]*Sydney Brenner Institute for Molecular Bioscience, Faculty of Health Sciences, University of the Witwatersrand, Johannesburg, South Africa;* [2]*National Institute of Biomedical Genomics, Kalyani, West Bengal, India*

Introduction

The majority of the human population resides in developing countries. Although there are several different classifications for "developing" countries, most countries in the broad geographic regions of Africa, South Asia, Southeast Asia, and South America are categorized as "developing." Consequently, these regions are often collectively referred to as the "developing world." In addition to the nonuniform distribution of populations across the globe, human genetic diversity is also strongly geographically skewed. Africa is not only the cradle of humankind but also the longest abode of our species, which has led to unparalleled ancestral, ethnolinguistic, and genetic diversity. By virtue of being on the route of the out of Africa migrations, regions such as South Asia also harbor considerably higher genetic diversity compared to other regions. In addition, many South American populations harbor different ancestries such as Native American, European, and African, due to the large-scale movement of people into this region over the second part of the last millennia.

Despite overall higher genetic and ancestral diversity, there is limited genomic data from populations inhabiting the developing world.[1] Moreover, much of our current knowledge of genomic diversity from these regions is based on studies conducted on small sample sizes and often based on a limited set of markers (e.g., only mitochondrial or Y chromosome markers or small genotyping arrays), which narrows the scope and utility of these datasets to contextualize new data. Furthermore, the absence of appropriate consent and limiting governance frameworks for sharing these datasets via global repositories have made some of the existing datasets inaccessible for future research. Sadly, in global genomics studies and databases, while data for populations from Europe, North America, and East Asia are increasing at an unprecedented pace, the relative representation of populations from the developing world is effectively diminishing. Besides inadequacy of infrastructure and capacity, lack of systematic investment lies at the center of the problem and unless urgent interventions are sought to reverse this trend, it is anticipated that the relative representation will reduce further in the coming years.

Over the last decade, large consortium-based studies have been the major source of genomic data generation. These studies include global projects aimed at assessing genomic diversity (such as the Human Genome Diversity Project (HGDP), HapMap, and 1000 Genomes Project), and they typically define populations by ethnicity/language and/or geography and have a small number (between 6 and 100) of study participants representing each population.[2–9] Although most of these studies have attempted to include a wide geographic representation, many broad regions (such as Southern and Northern Africa) and ancestries (two of the major African ethnolinguistic divisions; Nilo-Saharan and Afro-Asiatic) are completely absent in these datasets. For various reasons, the sampling strategies for studies such as the 1000 Genomes Project have not been optimal for some of the populations. For instance, most individuals representing South Asia have been sampled from either the US or the UK. Some major gaps are being addressed by geographic region-specific large consortia such as the Human Heredity and Health in Africa Consortium (H3Africa) and Genome Asia.[10,11] These consortia are discussed in detail later in this chapter.

Country/geographic region-level large genomic initiatives such as the Million Veteran Program (MVP), All of US, UK Biobank (UKBB), Japan Biobank (JBB), Dutch Genome Project, Qatar Biobank (QBB), FinGen, and Iceland's genome projects are becoming more and more frequent and are achieving increasingly larger participant sizes.[12–21] Although most of these studies have a broad public health focus, these large datasets have contributed immensely to the understanding of the peopling of these countries. Such efforts have been attempted in a few countries from the developing world and only in a handful of these projects (such as the initiatives in Mexico and Brazil) have reached the scale to be able to generate datasets that can inform population genomics in a substantial manner.[22,23]

A substantial fraction of the large genomic datasets is generated by disease/trait-focused consortia. However, by design, a majority of these are not optimized for population genetic insights.[24–26] Moreover, the governance framework of most of these does not permit access to individual-level data, which is essential for population genetic analysis. While there are efforts to make large consortia samples genetically and ancestrally diverse, these efforts generally take the form of including diaspora populations such as African Americans and US-Latinos. Therefore the representation of populations from continents such as Africa, South Asia, and South America is still noticeably low in most of these studies (Table 3.1).

Despite limited coverage and representation, the datasets from large consortia have been instrumental in advancing research and capacity building for population genetics in the developing world. In this chapter, we present examples of data from large consortia that have complemented existing population genetic datasets to expand our understanding of the ancestral composition and history of African populations. An independent section presents the contribution of such datasets to South Asian population genetics. Although South American populations also harbor high genomic diversity, due to limited space we have not delved into populations from these regions in the present chapter. We summarize widely used genomic resources that have originated from large consortium studies and then outline some of the key strengths and limitations of the secondary use of large consortium datasets for population genomic studies and inferences. Finally, we discuss the major factors that need to be considered and incorporated during the design/implementation of ongoing/future large consortium studies to enable the field of population genomics to benefit from the data of these health-focused initiatives.

Table 3.1 Representation of participants with African ancestry compared to participants from the African continent in selected large consortia.

Consortium name	Total number of participants	Percentage of participants with African ancestry (%)	Percentage of Africans from Africa (if >1%)
gnomAD	∼76,000	∼27	None
UK Biobank (UKBB)	∼500,000	∼1.50	None
TOPMed	∼180,000	∼29	None
Genotype-Tissue Expression (GTEx) Version 8	∼1000	∼13	None
Global Lipids Genetics Consortium (GLGC)	∼165,000	∼6	None

Large consortium studies have played a major role in assessing African diversity and population structure

Population splits dating back tens of thousands of years define the core population structure in Africa.[27] The high-level core structure of African populations includes four groups: (1) populations ancestral to present-day Khoe-San inhabiting Southern Africa, (2) populations ancestral to present-day rain forest foragers (RFF) inhabiting Central Africa, and distinct population groups inhabiting (3) East and (4) West Africa. Archaeological data and data from ancient genomes suggest that despite a deep split, these groups were not completely isolated and rather had prolonged periods of contact and gene flow between them,[28] which led to the establishment of geographically stratified ancestry clines. Post-agriculture migrations within the continent as well as the movement of people back into the continent over the last few millennia has further reorganized the pattern of peopling of the continent. This has led to a population structure that in addition to geography, is defined largely by these major ancestral divisions and the extent of gene flow between them.[27,29] This long and complex history has led to the development of over 2000 languages/ethnolinguistic groups[30] in Africa. Most of the African ethnolinguistic groups that have been included in genomic studies so far have shown the presence of distinctive ancestral compositions and unique genetic components.

Many standard population genetic studies[31–42] based on moderate sample sizes and genotype-array datasets have been major sources of data for African population genomics. These studies have been instrumental in providing a broad view of the African genetic landscape and remain an important source for generating insights into population structure, gene flow, and demographic history. However, over the last few years, large consortium studies have emerged as a key contributor to the comprehensive assessment of the extent of African diversity and a deeper understanding of population structure within specific geographic regions. Table 3.2 summarizes some of the key large consortium studies that have contributed substantially to population genomics in Africa.

The first major global consortium that had a substantial representation of sub-Saharan African (sSA) populations was the HapMap study. The first phase (HapMap 2005) of this study included the Yoruba from Nigeria (YRI), while the final phase (HapMap 2010) included two additional East

Table 3.2 Examples of large consortia that have contributed to African population genetics.

Consortium	Study	Sample size of African participants	Data type	Country
International HapMap 3 consortium (phase 3)	International HapMap 3 Consortium et al.[6]	338	Array	Kenya, Nigeria
1000 Genomes Project (phase 3)	Auton et al.[5]	~500	WGS	Nigeria, Nigeria, Kenya, Gambia, Sierra Leone
Simons Genome Diversity Panel (SGDP)	Mallick et al.[2]	48	WGS	Sudan, Nigeria, Nigeria, Western Sahara (Morocco), Gambia, Algeria, Senegal, Sierra Leone, Nigeria, Central African Republic, Congo, Kenya, Kenya, Kenya, Kenya, Namibia, South Africa, Botswana or Namibia, Cameroon, Kenya
Human Genome Diversity Project (HGDP)	Bergström et al.[3]	115	WGS	Algeria, Senegal, Nigeria, Central African Republic, Democratic Republic of Congo, Kenya, Namibia, South Africa
African Genome Variation Project (AGVP)	Gurdasani et al.[43]	3201 481	WGS Array	Uganda, Ethiopia, South Africa Uganda, Ethiopia, Kenya, South Africa, Nigeria, Ghana, Gambia
Uganda Genome Resource (UGR)	Gurdasani et al.[44]	20004 778	WGS Array	Uganda
Human Heredity & Health in Africa (H3Africa)	Choudhury et al.[10]	348	WGS	Burkina Faso, Mali, Nigeria, Ghana, Benin, Botswana, Zambia, Cameroon, Cameroon
H3Africa	Sengupta et al.[45]	~5000	Array	South Africa
TrypanoGEN	Mulindwa et al.[46]	159	WGS	Guinea, Ivory Coast, DRC, Uganda
NeuroGap	Atkinson et al.[47]	900	Array	Ethiopia, Kenya, South Africa, and Uganda

African populations; the Luhya from Kenya (LWK) and Masai from Kenya (MKK) and an African diaspora population from Southwest USA (ASW).[6,7,9] One of the major insights from this study was the demonstration of a much higher genetic diversity and lower linkage disequilibrium (shorter haplotype blocks) in African populations compared to non-Africans. The 1000 Genomes Project, in addition to the three African populations included in the HapMap study (YRI, LWK, and ASW), included three additional Central-West and West African populations (Esan in Nigeria (ESN), Gambian in Western Division—Mandinka (GWD), Mende in Sierra Leone (MSL)). It also includes one additional African diaspora group from the Caribbean islands (ACB).[4,5,8] Based on about 100 low-coverage whole genome sequences (WGS) from each population, this study demonstrated that African genomes harbor over 20% more genetic variants compared to the genomes from non-African populations.[5] The use of high-coverage WGS data in recently published Phase 4 of the 1000 Genomes Project enabled further refinement of estimates for the distribution of the structural variants and identification of genomic regions that are difficult to study by low-coverage sequencing.[4,48] However, the 1000 Genomes Project only included populations belonging to one (Niger—Congo speaking populations) of the several major African ethnolinguistic divisions. Moreover, the sampling of four of the five groups from West Africa limits the scope of use of this dataset for investigating the immense geographically stratified genomic diversity that exists in African populations.

The inclusion of high-depth WGS from a much more diverse set of African genomes, including the hunter-gatherer populations, in global initiatives such as Simons Genome Diversity Project[2] and the Human Genome Diversity Project[3] (HGDP) enabled the comparison of deep demographic histories of the major ethnolinguistic groups. These investigations estimate the ancestors of current-day African hunter-gatherers such as Khoe-San and the RFF to have had substantially larger population sizes for a large part of human history.[2] Although these datasets have been used extensively for contextualizing novel genomes, the representation of each group by only a handful of samples only allows for a small subset of analyses to be performed.

Insights from initial genomic surveys made it evident that global studies with limited coverage of African populations were not sufficient to provide a comprehensive view of the African genetic landscape. The African Genome Variation Project[43] (AGVP) was the first major Pan-African initiative aimed at characterizing the genomic diversity of African populations. This study, based on dense genotype-array data from 18 populations from East, West, and South Africa, showed that geography-specific admixture with Eurasian ancestry (originating from back-to-Africa migrations) and hunter-gatherer populations play a major role in differentiating present-day Africans. Moreover, it detected that the admixture patterns in many sub-Saharan African populations were complex and multilayered, with individuals receiving gene flow from more than one of these ancestries. Furthermore, the inferred dates of these admixture events ranged between a few hundred to several thousand years ago, providing evidence for several waves of migration within the continent as well as back migration into Africa. The analysis of 320 low-coverage WGS from population groups from Ethiopia, Uganda, and South Africa detected a large proportion of variants to be unique to individual groups. In-depth evaluations based on these sequences provided clues for the design of more efficient genotyping platforms and imputation resources for African populations.

The Human Heredity & Health in Africa (H3Africa) Consortium[49] was one of the first large-scale genomic initiatives in Africa. Although the projects included in the consortium are primarily focused on characterizing genetic architectures of communicable and noncommunicable diseases, the consortium dataset representing over 50,000 individuals, is the largest genomic dataset generated from the continent to date. The population genomic analysis of some of these datasets has been instrumental in the fine-scale characterization of admixture profiles as well as in suggesting the routes and timelines of the migration events that might have resulted in these admixtures. For instance, based on an analysis of around 500 WGS sequences, an H3Africa study proposed a potential route for Bantu-speaker migration to East and Southern Africa via Angola and Zambia.[10] Similarly, another study showed high genetic differentiation of Nilo-Saharan speaking populations[46] and demonstrates substantial genetic contributions from different Nilo-Saharan speaking groups to some of their neighboring Niger—Congo speaking populations.

A major strength of large consortium datasets is their ability to provide insights into the genetic landscape of a specific geographic region by the inclusion of several major population groups residing in an area. The Uganda Genome Resource[1,44] study genotyped over 6400 individuals from rural Uganda that contained a mix of groups from the Central region of Uganda (the Baganda, Basoga, and Batooro), populations from South-Western Uganda (Bakiga, Banyankole, and Bafumbira), and migrant populations from Rwanda, Burundi, and Tanzania (Banyarwanda, Rwandese Ugandans, Burundi, and Batanzania, respectively). Although most of these populations were Niger—Congo speakers, the three broad geography-based groups (Central-Ugandan, South-Western Uganda, and migrants) segregated into separate clades in fine-scale population structure analysis, each characterized by different levels of Eurasian and/or Nilo-Saharan gene flow, suggesting differential gene flow as a major source of the structure.

The H3Africa AWI-Gen study,[45] which included over 5000 individuals from South Africa, had representation of eight major South African Niger—Congo speaker groups. The genetic data from the study participants showed a very broad separation of the three major ethnolinguistic divisions (Nguni, Sotho-Tswana, and Tsonga). Similar to the observation of the impact of Eurasian gene flow in the UGR study, differential Khoe-San gene flow emerged as a key contributor to this population structure. The study also detected different levels of Khoe-San ancestry in participants who self-reported belonging to the same group when sampled from two different geographic regions: one corresponding to the area of linguistic majority and the other corresponding to regions where they have recently migrated to. Moreover, the consideration of parent and grandparent ethnolinguistic affiliations provided an assessment of the trends in gene flow between groups.

The NeuroGap Consortium study[47] based on a similar investigation into the relationship between language and genetics in populations from four African countries based on tracking language transmission through the pedigree, reported prominent shifts in frequency of languages over three generations. The recently reported 54-Genes study[50] also reports the existence of considerable genetic distances between many of the Nigerian ethnolinguistic groups.

Taken together, these studies/datasets demonstrate strong population structure in Africa, even at the level of individual countries, and highlight the critical importance of considering sampling geography and information about parent and grandparent language/ethnicity in addition to participant language/ethnicity, in ensuring the optimal genetic representation of a group. It would not have been possible to uncover the complex relationship between language, geography, and self-reported ethnolinguistic affiliation, without unbiased deep sampling from particular geographic regions, which is a core feature of large consortium studies.

Furthermore, the secondary analysis of datasets originating from several other consortium studies[51] has contributed considerably to our understanding of population structure and demographic history, not only at the level of the continent but also for specific geographic regions. Despite these efforts, less than 10% of African ethnolinguistic groups have representative genetic data and a far smaller percentage of these groups have data that are large enough for in-depth population genomic analysis. Even at the geographic level, large parts of the continent are un-represented or extremely underrepresented in the genomic data space. Therefore further large consortium initiatives are desperately needed to bridge these glaring gaps. Moreover, initiatives to integrate, harmonize, and conduct population genetics analyses on datasets that have already been generated by current consortia, could provide valuable information and insights.

Resources generated from large consortium datasets

In addition to contributing directly to population genetic studies, datasets (and tools) originating from large consortia also contribute to resources that are routinely used in genomic research (Fig. 3.1). The H3Africa consortium used a combination of novel WGS data and publicly available WGSs to develop a genotyping array that is enriched in African variants and selected based on linkage disequilibrium block variations in African populations (https://chipinfo.h3abionet.org/). The Africa-focused genotyping platform (known as the H3Africa array) has so far been used to genotype over 50,000 participants from sub-Saharan Africa, resulting in it being the most widely used genotyping platform for

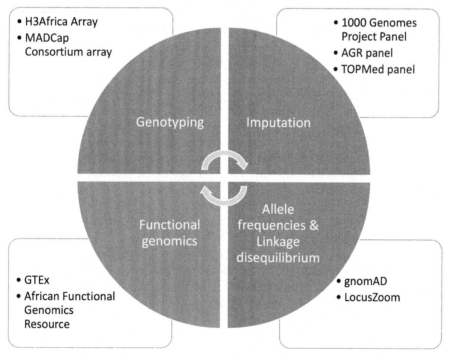

FIGURE 3.1

Examples of resources for African genomics that originate from consortium datasets. The central circle shows the categories.

continental African populations. Similarly, the Men of African Descent and Carcinoma of the Prostate (MADCaP) study has recently developed a genotyping array that is optimized for capturing African genotype diversity and improving imputation in African populations.[52] As several African WGS datasets have been generated since the development of these arrays, and due to plummeting cost of sequencing, a lot more data is expected to be generated soon. Further refinement of the existing arrays and development of other Africa-centric arrays can be anticipated in the near future.

Despite limited coverage, the 1000 Genomes panel continues to be one of the most widely used imputation panels for African Genome-Wide Association Study (GWAS) datasets. A major genomic resource originating from datasets generated by AGVP[43] and UGR[44] is the African Genome Resources (AGR) imputation panel, which is currently hosted at the Sanger Imputation Service. This platform by virtue of the inclusion of over 3000 African genomes from various parts of the continent has emerged as a widely used panel for imputing continental African genotype datasets. The imputation panel generated by the TOPMed consortium, although limited in continental African representation, is becoming very popular for imputing African datasets due to its size and inclusion of a substantial proportion of African diaspora genomes.[53] A recent study has shown these panels to be the best for imputing continental African datasets.[54] The novel sequence datasets that have been/are being generated by other consortia, both from Africans and the diaspora populations, have the potential to lead to the further development of more diverse, representative, and efficient imputation panels for imputing African genotype datasets.

The reference genome provides the core template for the identification of variants and the assembly of newly sequenced genomes. Studies over the last years have shown that due to large-scale differences in the genome architecture, the current linear reference genome does not provide comprehensive coverage for all populations. Especially populations from Africa and those harboring African ancestry often show suboptimal assembly. Two large consortium initiatives are underway to address this by generating a more in-depth representation of the genome: (1) The Human Pangenome Reference Consortium (HPRC)[55] is building a comprehensive reference genome by employing graph-based, telomere-to-telomere representation of the diversity that exists in the human genomes from different parts of the world; (2) The Genome Reference Consortium (GRC) (https://www.ncbi.nlm.nih.gov/grc) is employing orthogonal approaches to bridge the existing gaps in the human genome that includes generating alternative assemblies of structurally variant loci in the genome. Both these efforts are attempting to ensure an optimal representation of African genomic diversity in their reference genome/genome graphs.

Similar limitations have also been noted for the performance of current recombination maps, particularly for African population groups such as Khoe-San[56] that are missing in current reference datasets. The use of recombination maps augmented by a small set of representative genomes from this group was found to strongly impact population genetic estimates, suggesting that the generation of more elaborate recombination maps using local reference datasets could be a useful way to improve the resolution of genomic datasets from the developing world.

There are several other frequently used resources generated from large consortium datasets that have been extremely valuable for genomics research on populations from the developing world. These include allele frequency and linkage disequilibrium (LD) architecture resources such as the 1000 Genomes Browser,[5] gnomAD,[57] LocusZoom,[58] and resources that integrate other omics data, such as GTEx,[59] African functional genomics resource (https://github.com/smontgomlab/AFGR).

Datasets based on diaspora populations residing in countries like the US (e.g., All of US project is enriched for minority groups) and the UK (e.g., UK Biobank with $\sim$8000 participants with African ancestry) could also have enormous value as genomic resources. For example, the dataset from the TOPMed consortium, which includes approximately 30,000 African genomes (predominantly from the African diaspora in the US), has been made available via the TOPMed imputation service and has become a widely used resource for imputing datasets from African ancestry populations. Augmenting the representation of continental African populations in these resources could substantially increase the applicability and efficiency of these resources. Data Science for Health Discovery and Innovation in Africa (DS-I Africa) (https://dsi-africa.org/), a new NIH-funded initiative, is underway to boost the generation of genomics resources utilizing existing and novel African datasets.

Insights into the genetics of South Asians from large consortium studies

South Asia is a subregion of Asia that is broadly defined by the Indian subcontinent and includes India, Pakistan, Bangladesh, Nepal, Bhutan, Sri Lanka, and Maldives. Despite being one of the most densely populated geographic regions, populations from this region are disproportionately underrepresented in genomic studies. One of the earliest efforts to catalog and interpret worldwide human variation, including South Asian participants, was by Rosenberg et al.[60] using the HGDP panel, where they analyzed 377 autosomal microsatellite loci in 1056 individuals from 52 diverse global populations. An unsupervised clustering algorithm, used to identify population structure in the data, identified five major genetic clusters in the worldwide data that correlated strongly with the geographical locations of extant populations. The major clusters that were identified were: Africa, Eurasia, East Asia, Oceania, and the Americas. Interestingly, this approach did not separate populations from Europe, the Middle East, and Central South Asia from each other (Fig. 1 in Ref. 60). Six years later, a follow-up study using many of the same individuals, but now with a much larger array of 650,000 common single nucleotide polymorphisms (SNPs), revealed seven clusters.[61] This study identified and differentiated ancestries specific to the Middle East (West Asia) and Central South Asia, thus teasing them apart as separate clusters from the other major populations of the world (Fig. 1 in Ref. 60). What this analysis also revealed is that all the West Asian (Middle Eastern) populations are significantly admixed, primarily with large genetic inputs from the European and Central South Asian populations. Despite the uniqueness of these ancestries, the global genomics consortia had very limited coverage of these Asian geographies and ancestries. The HapMap and 1000 Genome Project Phase 1 study did not have a representative population from West Asia and included only one South Asian population (Gujarati Indian from Houston [GIH]).[6,8]

Initial population genetic studies[62] estimated that all extant Indian populations were derived from admixture between two ancestral populations, with the contribution of the Ancestral North Indian (ANI) ancestry being higher among extant north Indian populations and that of Ancestral South Indian (ASI) being higher among extant south Indian populations. The study by Reich and colleagues,[62] using over 500,000 biallelic autosomal SNPs, also found a north-to-south gradient of genetic proximity of Indian populations to western Eurasians/Central Asians. In a follow-up study,[63] these investigators showed that between 1900 and 4200 years before present (ybp), there was extensive admixture among

Indian population groups, followed by a shift to endogamy. A more recent study[64] analyzing data on more than 800,000 SNPs on 367 individuals from 18 mainland and two island populations identified two more ancestral components in the mainland of the Indian subcontinent, besides the ANI and ASI. These two newly identified ancestral components are primarily seen among Tibeto-Burman (TB) speaking tribal populations of northeast India and Austro-Asiatic (AA) speaking tribal populations of Central and East India, and the authors identified them as Ancestral Austro-Asiatic (AAA) and Ancestral Tibeto-Burman (ATB).

Although the 1000 Genomes Project Phase 3 had four other representative populations for South Asia, they all broadly correspond to just one of the South Asian Ancestry components (ANI), resulting in much of the genetic diversity from this geographic region remaining underrepresented in global databases.[65] Despite limited coverage, this study detected more variants per genome in the South Asian Bengalis compared to European to East Asian populations.[5] Although national level projects such as the Indian Genome Variation Consortium (2005)[66] and the more recent Indian Exome Reference Database[67] have been initiated, they focus on narrow geographic regions and did not reach the sample sizes required for in-depth characterization of genetic diversity and demographic history, resulting in significant gaps in the estimation of genetic diversity from this region.

To address this underrepresentation of Asian populations in human genetic studies, a consortium to catalog population-specific reference genome datasets for Asian populations was undertaken by the GenomeAsia 100K Project. This project, in its first publication,[11] describes itself as a pilot phase of a bigger project which would catalog genome diversity in Asian populations. Its aim is to identify novel variants and utilize LD and population structure information to design a new genotype chip to facilitate and enhance genome-wide studies of diseases in these populations. The consortium analyzed a WGS reference dataset from 1739 individuals from 219 population groups and 64 countries across Asia. The dataset includes 598 sequences from India, 156 from Malaysia, 152 from South Korea, 113 from Pakistan, 100 from Mongolia, 70 from China, 70 from Papua New Guinea, 68 from Indonesia, 52 from the Philippines, 35 from Japan, and 32 from Russia.

This large consortium dataset replicated the population genetics findings in previous studies done in India, Malaysia, and Indonesia, and clearly showed that these countries were populated and inhabited by multiple ancestral populations and that the extant populations are the result of intricate and complex admixture events throughout their history. In a first of its kind, this study also described the variation in archaic admixture across different populations living in Asia. Besides delineating the population genetics of the region, this study also emphasized medically relevant applications to determine how the results of larger ongoing GenomeAsia studies could be used to improve human health. Additionally, this study identified an overall 23% of protein-altering variants, of which a large number were novel ($n = 194,585$) and had minor allele frequencies $\geq 0.1\%$ in the GenomeAsia dataset. The study could also identify the presence of a large number of endogamous Asian populations in virtually every geographical region where the study was conducted. Moreover, recent studies that are based on or supported by this dataset (e.g., Tagore et al.[68]) are providing deeper insights into the population genetics of specific populations and/or geographic regions. The GenomeAsia dataset has been instrumental in bridging the gaps in the representation of Asian genomic diversity in the global reference genome panel. A panel based on this dataset can be employed for imputation via the Michigan Imputation Server (https://imputationserver.sph.umich.edu/).[69]

Use of large consortia datasets for population genomics inferences—the promises and challenges

One of the major features of large consortium studies is a relatively larger sample size compared to standard population genomic studies, which confers several advantages for population genetics investigation (Fig. 3.2). For instance, the lowest ballpark representative sample size for a population in studies such as 1000 Genome Project and AGVP was generally around a 100 individuals, which provides a relatively robust estimate of common variant allele frequencies compared to smaller population genetic studies. The consortium studies designed for GWASs typically have much larger sample sizes, which often go up to several thousand participants, enabling robust estimation of distribution and frequencies of rare variants.

A considerable proportion of variants that have been shown to have clear biological relevance (such as American College of Medical Genetic and certain PharmGKB variants) tend to occur at very low frequencies in a population and are often not included in or efficiently genotyped by genotyping arrays. Even when these are imputed, due to their low frequency, the imputation scores are often low and therefore many of these tend to get removed by score-based filters. Their reliable identification and

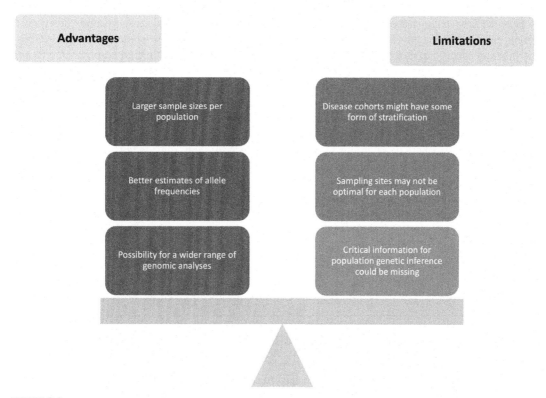

FIGURE 3.2

Advantages and limitations of using large consortium datasets for population genetic inferences.

comparison of frequencies between populations are only possible using whole genome/whole exome sequencing approaches, which are often made possible by large consortium datasets. For instance, recently a study[70] using the H3Africa consortium WGS data surveyed the distribution of some of the pharmacogenetically relevant variants in the G6PD gene in several African populations and demonstrated fine-scale differences in the distribution of variants not only at the continental level but also within different African regions.

Sample sizes also play a key role in our ability to identify signatures of selection from genome-scale datasets. Some of the estimates, such as Wright's Fixation Index (FST) and Population Branch Statistic (PBS) that are intrinsically based on allele frequencies, require a minimum sample size to robustly identify signals and therefore become less reliable as the sample sizes decrease.[71] Even other approaches, such as haplotype homozygosity-based selection scans, that are not inherently based on allele frequencies, employ frequency estimates for filtering. For instance, the widely used estimate, integrated haplotype homozygosity score (iHS), employs a minor allele frequency threshold of 0.05.[72] Therefore for studies based on very small sample sizes, selection scans performed using such approaches become less reliable. Sample sizes are also critical to several other population genetic estimates, including admixture dating, admixture mapping, and demographic history inference. Therefore large consortium studies are generally better equipped to generate more robust estimates of population history, migration, and gene flow.

However, the large consortium studies, especially those focused on disease genetics, also have certain constraints that might limit their utility for population genetic inferences. Genetic studies investigating specific diseases often recruit participants belonging to multiple ethnolinguistic groups from a single study site. In many settings, such as Africa, geography is an important defining component of an ethnolinguistic group. As self-identified ethnicity is complex and strongly influenced by cultural and social factors, the groups that are sampled from regions that are not their historical area of residence might not be represented optimally in such studies. Studies such as Sengupta et al. (2021) have shown that the same ethnolinguistic group, when sampled from their area of majority origin compared to an area where these groups have recently moved to, show considerably different ancestral composition.[45] Therefore population genetic analyses of data originating from consortium studies need to take the sampling geography into account and focus their inferences only on those groups that can be expected to be reliably represented at a particular sampling site. This is particularly important in urban centers where a large proportion of residents could be recent migrants with different ethnolinguistic affiliations and gene flow between groups is often common. In addition, in case-control GWAS datasets, the participants recruited as cases and controls may have different proportions of ethnolinguistic groups and might have distinctive allele frequencies for several variants that may be associated with the disease/trait under investigation. Therefore caution needs to be exercised in employing such datasets for analyses such as allele frequency estimation or selection scans.

Strategies to enable large consortium studies to support population genomics research

Datasets originating from large disease-focused consortium studies have the potential to make important contributions to understanding ancestry and admixture in populations from the developing world. Unfortunately, due to a variety of reasons, a substantial portion of these datasets are not available/accessible for population genetic inferences. In light of the acute shortage of genomic data and the absence of national/regional genomic initiatives in developing countries, the inability to harness these existing datasets and resources presents a major barrier to population genomics research. In the following paragraphs, we outline some of the major factors that have either prohibited or restricted the use of data from these studies for population genetics research (Fig. 3.3). Moreover, we have recommended strategies that ongoing and future consortium studies might incorporate to enable a more multifaceted utilization of the datasets generated by these initiatives (Fig. 3.3).

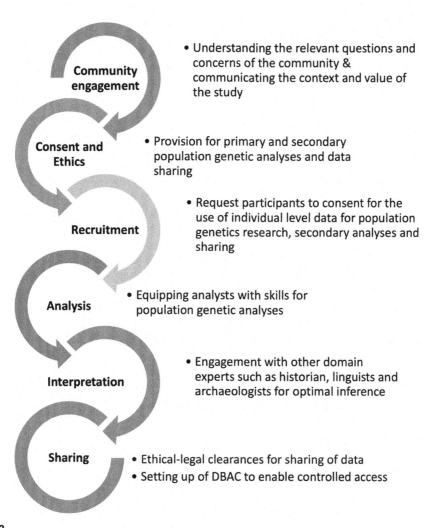

Community engagement
- Understanding the relevant questions and concerns of the community & communicating the context and value of the study

Consent and Ethics
- Provision for primary and secondary population genetic analyses and data sharing

Recruitment
- Request participants to consent for the use of individual level data for population genetics research, secondary analyses and sharing

Analysis
- Equipping analysts with skills for population genetic analyses

Interpretation
- Engagement with other domain experts such as historian, linguists and archaeologists for optimal inference

Sharing
- Ethical-legal clearances for sharing of data
- Setting up of DBAC to enable controlled access

FIGURE 3.3

Provisions at each stage of a study to enable optimal population genetic inferences from large consortia datasets.

First, many disease genetics studies do not include any population genetics questions in their core objectives. Even for the studies that conduct such analyses, the consent and ethics approvals often do not explicitly include the provision of sharing for secondary use of individual-level data. Therefore while such datasets (e.g., GWAS summary statistics) continue to contribute to research on the genetics of specific diseases or traits, similar use of these datasets in population genetic studies is precluded. This also precludes the use of the data to investigate new approaches and algorithms, to act as reference datasets (such as local ancestry panels and imputation panels), or their combined analysis with other datasets. Reanalysis of some of these datasets with newer tools and resources might provide a better interpretation, compared to what was possible at the time of the original study. Conversely, if available, these datasets could also be valuable for contextualizing newer datasets. Therefore for future disease-consortium studies, it is important to include at least some provision in the consent to enable population genetic inference and secondary use of these datasets. To balance this with the need for participant privacy, a strong governance framework needs to be established that would ensure that the future use of this data adheres to the specified purpose and does not place the individuals or the communities that have contributed to the biological samples at risk.

Second, in many cases, studies that have consented to the participants for sharing and secondary use of data, the respective questionnaires did not include questions that capture all the essential parameters that are required for a comprehensive population genomic analysis. This is especially relevant in settings such as Africa, where ancestry relates to a complex interplay of factors such as geography, language, and ethnolinguistic affiliation. As noted above, in such settings, in addition to information on the participants, information on their parents and grandparents could also be useful in the proper categorization of individuals into groups. Therefore it is important that future studies design their questionnaires to capture the minimum information necessary for population genetic analyses.

It is also crucial to tailor the questions and inputs to the context of the specific populations that are being studied. For instance, in a South Asian setting, caste might be an important determinant/covariate of ancestry and should be asked in a culturally sensitive manner. Similarly, for populations that have been reshaped by recent large-scale migrations, information on origin or place of birth, and migration history would be valuable. Also, in some cases due to historical and social factors, certain questions which are routine for one setting might result in discomfort or raise concerns in other communities. For example, the communities that have been impacted by recent tension among constituent groups, the use of language or ethnolinguistic identity as a classifier should be avoided. The researchers analyzing the data need to ensure that extreme caution is exercised when reporting the outcomes of sensitive information for the study. Therefore continuous and in-depth engagement with each community as well as involving domain experts, such as sociologists and historians, as well as other stakeholders is paramount to planning the optimal strategy to balance the prerequisites for genomic research with the interest, contentions, and history of the community.

Finally, despite all the abovementioned strategies, there might be cases where sharing of individual-level data might not be possible either due to the sensitivity of the study or other constraints. In those cases, it would be immensely valuable if the large consortium studies could share some summary-level statistics that can be employed in population genetics studies. These might include principal component loadings and admixture projections with respect to index populations from the 1000 Genomes Project and other public datasets that could allow for ancestry deconvolution.

Conclusions and future directions

Large consortium projects, such as the 1000 Genomes Project and the HGDP, have played a pivotal role in generating data to inform the contours of human genetic diversity and ancestral composition at a global scale. Although they had limited representation of the developing world, these studies and the ease of accessing these datasets, have made them core datasets for genomics research and a major resource for genomics capacity development. Unfortunately, several consortia that had a primary focus on population genomics have reached or will soon reach completion. From a developing world population's genomics perspective, it is critical that such global initiatives continue to receive support, get scaled, and become focused on major population gaps that are yet to be addressed.

Geography-specific large consortium studies such as the AGVP, H3Africa, and GenomeAsia have been instrumental in shaping our current understanding of the peoples of specific geographic regions. Their datasets, due to their accessibility, have been a crucial baseline and have been widely used for contextualizing other datasets originating from the corresponding geographic regions. They have also been instrumental in the development of core genomic resources such as genotyping arrays and reference panels (e.g., imputation and inferring local ancestry). As the strength of these resources depends critically on the size of the datasets (for instance, reference panels require WGS data on thousands of individuals), many of these might not have been possible, or at least, as comprehensive, without the resources and infrastructure of large consortium studies. Some of these large consortia have also been representing the developing world in ongoing global initiatives such as the Human Pangenome Reference Consortium and geographic region-specific functional genomics initiatives.

Population genomics in the developing world is strongly impacted by the scarcity of support at the country level. Country-specific initiatives such as UKBB, JBB, and QBB have been key to providing larger and more fine-scale data in developed countries and usually receive generous government support, at least when they are initiated. Given the promises demonstrated by these studies, existing national initiatives need to be scaled up and new studies need to be initiated to represent more countries in the developing world. Since the limited budget for scientific research and the unavailability of infrastructure and human resources to conduct such studies are often major challenges, discussions between national governments and contributions from major international funders are needed for such studies to be realized. The development of infrastructure and capacity would not only ensure that populations inhabiting these countries are able to benefit from innovations in genomic research, but also equip local researchers to address health challenges that might be specifically important for a particular country.

Large consortium studies aimed at studying diseases have also yielded valuable datasets for population genetic inferences. In several cases, publications from the consortia or secondary studies utilizing these datasets have made substantial additions to the current understanding of the genomic diversity and history of specific countries. However, as population genetics has not been a focus during the conception and design of these studies, many such studies lack the permissions for secondary use as well as the information for making these datasets fully amenable to population genetic investigations. Given that such large-scale studies, under the auspices of initiatives, such as the African Population Cohorts Consortium (APCC) (currently in its formative phase) and the International Health Cohorts Consortium (IHCC) could be a major source of genomic data from most parts of the developing world, it is imperative that mechanisms are set up to enable researchers to harness these datasets for generating a better understanding of the overall genomic makeup of these populations.

Large consortium studies need to implement provisions to make individual level genotype/sequence data available from global repositories to enhance an in-depth understanding of populations living in the developing world. Moreover, these datasets need to be supported by minimal data on population genetics attributes as well as a mechanism so that bonafide researchers working on these populations are able to access and use the data in population genomic research. Similarly, the sharing of the ethnolinguistic group or country-level allele frequencies to databases, such as gnomAD and CLINVAR, could also be extremely valuable for future population genetics research. The population genetics community should also work on identifying and prioritizing essential data features and common standards to enable large consortium studies to collect and report relevant information. Moreover, the development of repositories for storage and tools for seamless integration of data into future population genetic studies would enhance the understanding of population affinities and how evolutionary processes have shaped genomic diversity.

References

1. Fatumo S, Chikowore T, Choudhury A, Ayub M, Martin AR, Kuchenbaecker K. A roadmap to increase diversity in genomic studies. *Nat Med*. 2022;28(2):243−250.
2. Mallick S, Li H, Lipson M, et al. The Simons genome diversity project: 300 genomes from 142 diverse populations. *Nature*. 2016;538(7624):201−206.
3. Bergström A, McCarthy SA, Hui R, et al. Insights into human genetic variation and population history from 929 diverse genomes. *Science*. 2020;367(6484). https://doi.org/10.1126/science.aay5012.
4. Byrska-Bishop M, Evani US, Zhao X, et al. High-coverage whole-genome sequencing of the expanded 1000 Genomes Project cohort including 602 trios. *Cell*. 2022;185(18).
5. Auton A, Abecasis GR, Altshuler DM, et al. A global reference for human genetic variation. *Nature*. 2015; 526(7571):68−74.
6. Altshuler DM, Gibbs RA, Peltonen L, et al. Integrating common and rare genetic variation in diverse human populations. *Nature*. 2010;467(7311):52−58.
7. A haplotype map of the human genome. *Nature*. 2005;437(7063):1299−1320.
8. 1000 Genomes Project An integrated map of genetic variation from 1,092 human genomes. *Nature*. 2012; 491:56−65.
9. Frazer KA, Ballinger DG, Cox DR, et al. A second generation human haplotype map of over 3.1 million SNPs. *Nature*. 2007;449(7164):851−861. https://doi.org/10.1038/nature06258.
10. Choudhury A, Aron S, Botigué LR, et al. High-depth African genomes inform human migration and health. *Nature*. 2020;586(7831):741−748.
11. Wall JD, Stawiski EW, Ratan A, et al. The GenomeAsia 100K Project enables genetic discoveries across Asia. *Nature*. 2019;576(7785):106−111.
12. Gaziano JM, Concato J, Brophy M, et al. Million Veteran Program: a mega-biobank to study genetic influences on health and disease. *J Clin Epidemiol*. 2016;70:214−223.
13. Ramirez AH, Sulieman L, Schlueter DJ, et al. The all of us research program: data quality, utility, and diversity. *Patterns*. 2022;3(8):100570. https://doi.org/10.1016/j.patter.2022.100570.
14. Sudlow C, Gallacher J, Allen N, et al. UK biobank: an open access resource for identifying the causes of a wide range of complex diseases of middle and old age. *PLoS Med*. 2015;12(3).

15. Nagai A, Hirata M, Kamatani Y, et al. Overview of the BioBank Japan project: study design and profile. *J Epidemiol*. 2017;27(3):S2−S8.
16. Boomsma DI, Wijmenga C, Slagboom EP, et al. The genome of The Netherlands: design, and project goals. *Eur J Hum Genet*. 2014;22(2):221−227.
17. Al Thani A, Fthenou E, Paparrodopoulos S, et al. Qatar Biobank cohort study: study design and first results. *Am J Epidemiol*. 2019;188(8):1420−1433.
18. Mbarek H, Devadoss Gandhi G, Selvaraj S, et al. Qatar genome: insights on genomics from the Middle East. *Hum Mutat*. 2022;43(4):499−510.
19. Gudbjartsson DF, Helgason H, Gudjonsson SA, et al. Large-scale whole-genome sequencing of the Icelandic population. *Nat Genet*. 2015;47(5):435−444.
20. Kurki MI, Karjalainen J, Palta P, et al. FinnGen provides genetic insights from a well-phenotyped isolated population. *Nature*. 2023;613(7944):508−518.
21. Zhou W, Kanai M, Wu K-HH, et al. Global Biobank meta-analysis initiative: powering genetic discovery across human disease. *Cell Genomics*. 2022;2(10):100192.
22. Sohail M, Chong AY, Quinto-Cortes CD, et al. Nationwide genomic biobank in Mexico unravels demographic history and complex trait architecture from 6,057 individuals. *bioRxiv*. 2022. https://doi.org/10.1101/2022.07.11.499652.
23. Patrinos GP, Pasparakis E, Koiliari E, et al. Roadmap for establishing large-scale genomic medicine initiatives in low- and middle-income countries. *Am J Hum Genet*. 2020;107(4):589−595.
24. Graham SE, Clarke SL, Wu K-HH, et al. The power of genetic diversity in genome-wide association studies of lipids. *Nature*. 2021;600(7890):675−679.
25. Mishra A, Malik R, Hachiya T, et al. Stroke genetics informs drug discovery and risk prediction across ancestries. *Nature*. 2022;611(7934):115−123.
26. Suzuki K, Hatzikotoulas K, Southam L, et al. Multi-ancestry genome-wide study in >2.5 million individuals reveals heterogeneity in mechanistic pathways of type 2 diabetes and complications. *medRxiv*. 2023. https://doi.org/10.1101/2023.03.31.23287839.
27. Schlebusch CM, Jakobsson M. Tales of human migration, admixture, and selection in Africa. *Annu Rev Genom Hum Genet*. 2018;19(1):405−428.
28. Lipson M, Sawchuk EA, Thompson JC, et al. Ancient DNA and deep population structure in sub-Saharan African foragers. *Nature*. 2022;603(7900):290−296.
29. Pfennig A, Petersen LN, Kachambwa P, Lachance J, Eyre-Walker A. Evolutionary genetics and admixture in African populations. *Genome Biology and Evolution*. 2023;15(4). https://doi.org/10.1093/gbe/evad054.
30. Omotoso OE, Teibo JO, Atiba FA, Oladimeji T, Adebesin AO, Babalghith AO. Bridging the genomic data gap in Africa: Implications for global disease burdens. *Globalization*. 2022;18(1):103.
31. Schlebusch CM, Skoglund P, Sjödin P, et al. Genomic variation in seven Khoe-San groups reveals adaptation and complex African history. *Science*. 2012;338(6105):374−379.
32. Henn BM, Gignoux CR, Jobin M, et al. Hunter-gatherer genomic diversity suggests a southern African origin for modern humans. *Proc Natl Acad Sci U S A*. 2011;108(13):5154−5162.
33. Tishkoff SA, Reed FA, Friedlaender FR, et al. The genetic structure and history of Africans and African Americans. *Science*. 2009;324(5930):1035−1044.
34. Lachance J, Vernot B, Elbers CC, et al. Evolutionary history and adaptation from high-coverage whole-genome sequences of diverse African hunter-gatherers. *Cell*. 2012;150(3):457−469.
35. Tishkoff SA, Gonder MK, Henn BM, et al. History of click-speaking populations of Africa inferred from mtDNA and Y chromosome genetic variation. *Mol Biol Evol*. 2007;24(10):2180−2195.
36. Barbieri C, Vicente M, Rocha J, Mpoloka SW, Stoneking M, Pakendorf B. Ancient substructure in early mtDNA Lineages of southern Africa. *Am J Hum Genet*. 2013;92(2):285−292.

37. Pagani L, Schiffels S, Gurdasani D, et al. Tracing the route of modern humans out of Africa by using 225 human genome sequences from Ethiopians and Egyptians. *Am J Hum Genet.* 2015;96(6):986–991.

38. Haber M, Mezzavilla M, Bergström A, et al. Chad genetic diversity reveals an African history marked by multiple Holocene Eurasian migrations. *Am J Hum Genet.* 2016;99(6):1316–1324.

39. Hollfelder N, Schlebusch CM, Günther T, et al. Northeast African genomic variation shaped by the continuity of indigenous groups and Eurasian migrations. *PLoS Genet.* 2017;13(8). e1006976.

40. Uren C, Kim M, Martin AR, et al. Fine-scale human population structure in Southern Africa reflects eco-geographic boundaries. *Genetics.* 2016;204(1):303–314.

41. Semo A, Gayà-Vidal M, Fortes-Lima C, et al. Along the Indian Ocean Coast: genomic variation in Mozambique provides new insights into the Bantu Expansion. *Mol Biol Evol.* 2020;37(2):406–416.

42. Patin E, Lopez M, Grollemund R, et al. Dispersals and genetic adaptation of Bantu-speaking populations in Africa and North America. *Science.* 2017;356(6337):543–546.

43. Gurdasani D, Carstensen T, Tekola-Ayele F, et al. The African genome variation project shapes medical genetics in Africa. *Nature.* 2015;517(7534):327–332.

44. Gurdasani D, Carstensen T, Fatumo S, et al. Uganda genome resource enables insights into population history and genomic discovery in Africa. *Cell.* 2019;179(4), 984-1002.e36.

45. Sengupta D, Choudhury A, Fortes-Lima C, et al. Genetic substructure and complex demographic history of South African Bantu speakers. *Nat Commun.* 2021;12(1). https://doi.org/10.1038/s41467-021-22207-y.

46. Mulindwa J, Noyes H, Ilboudo H, et al. High levels of genetic diversity within Nilo-Saharan populations: implications for human adaptation. *Am J Hum Genet.* 2020;107(3):473–486. https://doi.org/10.1016/j.ajhg.2020.07.007.

47. Atkinson EG, Dalvie S, Pichkar Y, et al. Genetic structure correlates with ethnolinguistic diversity in eastern and southern Africa. *Am J Hum Genet.* 2022;109(9):1667–1679.

48. Hanchard NA, Choudhury A. 1000 Genomes Project phase 4: the gift that keeps on giving. *Cell.* 2022;185(18):3286–3289. https://doi.org/10.1016/j.cell.2022.08.001.

49. Rotimi C, Abayomi A, Abimiku A, et al. Research capacity. Enabling the genomic revolution in Africa. *Science.* 2014;344(6190):1346–1348.

50. Joshi E, Biddanda A, Popoola J, et al. Whole-genome sequencing across 449 samples spanning 47 ethno-linguistic groups provides insights into genetic diversity in Nigeria. *bioRxiv.* 2022. https://doi.org/10.1101/2022.12.09.519178.

51. Busby GBJ, Band G, Si Le Q, et al. Admixture into and within sub-Saharan Africa. *Elife.* 2016;5(2016). https://doi.org/10.7554/elife.15266.

52. Harlemon M, Ajayi O, Kachambwa P, et al. A custom genotyping array reveals population-level heterogeneity for the genetic risks of prostate cancer and other cancers in Africa. *Cancer Res.* 2020;80(13):2956–2966.

53. Taliun D, Harris DN, Kessler MD, et al. Sequencing of 53,831 diverse genomes from the NHLBI TOPMed program. *Nature.* 2021;590(7845):290–299.

54. Sengupta D, Botha G, Meintjes A, et al. Performance and accuracy evaluation of reference panels for genotype imputation in sub-Saharan African populations. *Cell Genomics.* 2023;3(6).

55. Liao W-W, Asri M, Ebler J, et al. A draft human pangenome reference. *Nature.* 2023;617(7960):312–324.

56. van Eede G, Uren C, Pless E, et al. The recombination landscape of the Khoe-San likely represents the upper limits of recombination divergence in humans. *Genome Biol.* 2022;23(1). https://doi.org/10.1186/s13059-022-02744-5.

57. Karczewski KJ, Francioli LC, Tiao G, et al. The mutational constraint spectrum quantified from variation in 141,456 humans. *Nature.* 2020;581(7809):434–443.

58. Pruim RJ, Welch RP, Sanna S, et al. LocusZoom: regional visualization of genome-wide association scan results. *Bioinformatics.* 2010;26(18):2336–2337.

59. Aguet F, Barbeira AN, Bonazzola R, et al. The GTEx Consortium atlas of genetic regulatory effects across human tissues. *Science*. 2020;369(6509):1318−1330.

60. Rosenberg NA, Pritchard JK, Weber JL, et al. Genetic structure of human populations. *Science*. 2002; 298(5602):2381−2385.

61. Li JZ, Absher DM, Tang H, et al. Worldwide human relationships inferred from genome-wide patterns of variation. *Science*. 2008;319(5866):1100−1104.

62. Reich D, Thangaraj K, Patterson N, Price AL, Singh L. Reconstructing Indian population history. *Nature*. 2009;461(7263):489−494. https://doi.org/10.1038/nature08365.

63. Moorjani P, Thangaraj K, Patterson N, et al. Genetic evidence for recent population Mixture in India. *Am J Hum Genet*. 2013;93(3):422−438. https://doi.org/10.1016/j.ajhg.2013.07.006.

64. Basu A, Sarkar-Roy N, Majumder PP. Genomic reconstruction of the history of extant populations of India reveals five distinct ancestral components and a complex structure. *Proc Natl Acad Sci U S A*. 2016;113(6): 1594−1599.

65. Sengupta D, Choudhury A, Basu A, Ramsay M. Population stratification and underrepresentation of Indian subcontinent genetic diversity in the 1000 genomes project dataset. *Genome Biol Evol*. 2016;8(11): 3460−3470.

66. Brahmachari SK, Singh L, Sharma A, et al. The Indian genome variation database (IGVdb): a project overview. *Hum Genet*. 2005;118(1):1−11.

67. Vidhya V, More RP, Viswanath B, et al. INDEX-Db: the Indian exome reference database (phase I). *J Comput Biol*. 2019;26(3):225−234.

68. Tagore D, Aghakhanian F, Naidu R, Phipps ME, Basu A. Insights into the demographic history of Asia from common ancestry and admixture in the genomic landscape of present-day Austroasiatic speakers. *BMC Biol*. 2021;19(1). https://doi.org/10.1186/s12915-021-00981-x.

69. Das S, Forer L, Schönherr S, et al. Next-generation genotype imputation service and methods. *Nat Genet*. 2016;48(10):1284−1287.

70. Da Rocha JE, Othman H, Botha G, et al. The extent and impact of variation in ADME genes in sub-Saharan African populations. *Front Pharmacol*. 2021;12:634016.

71. Vitti JJ, Grossman SR, Sabeti PC. Detecting natural selection in genomic data. *Annu Rev Genet*. 2013;47(1): 97−120.

72. Sabeti PC, Varilly P, Fry B, et al. Genome-wide detection and characterization of positive selection in human populations. *Nature*. 2007;449(7164):913−918.

Implementing genomics research in developing countries: Common challenges, and emerging solutions

Austin W. Reynolds[1], Amanda J. Lea[2] and Maria A. Nieves-Colón[3]

[1]*Department of Microbiology, Immunology, and Genetics, School of Biomedical Sciences, University of North Texas Health Science Center, Fort Worth, TX, United States;* [2]*Department of Biological Sciences, Vanderbilt University, Nashville, TN, United States;* [3]*Department of Anthropology, University of Minnesota Twin Cities, Minneapolis, MN, United States*

Introduction

The completion of the Human Genome Project nearly 20 years ago and the resulting decrease in the cost of genomic sequencing invigorated many areas of genetic research. Our understanding of human evolution, history, and migration has grown immensely with this influx of data—from refining or rewriting existing models of human population history,[1–3] to uncovering evidence of admixture with archaic hominins,[4,5] to the discovery of entirely new lineages of the human family tree.[6] Our understanding of human—environment interactions has also improved as a result of genomic datasets, with further evidence for local genetic adaptations in human populations, and a better understanding of how our genotypes are modulated by development, lifestyle, and social factors.[7–9]

In parallel with these anthropological and evolutionary genetic advances, genome-wide association studies have uncovered thousands of single nucleotide variants (SNVs) associated with common diseases, providing key insights into the biological pathways and genes relevant to our health. They have also led to the development of a new field of research, known as precision medicine, that is centered on finding individualized diagnostic and treatment options for patients catered toward their particular genetic predispositions. This research has also spurred the development of polygenic risk scores (PRS)—based on the idea that complex diseases are determined by combined risk across many loci—and has provided new excitement for personalized risk assessment of complex conditions such as heart disease, type-2 diabetes, cancer, and psychiatric conditions.[10] PRS are already predicting many conditions more accurately than previous clinical models.[11–13] Importantly, these developments have only been made possible with the large-scale integration of electronic health records, and phenotypic and genomic data being done by projects such as the UK Biobank and the All of Us research program.

Population Genomics in the Developing World. https://doi.org/10.1016/B978-0-443-18546-5.00004-8

But despite these massive achievements, the genomic datasets used to produce them continue to be dominated by participants of European ancestry. A 2016 study found that over 80% of genomic studies were conducted in populations of European ancestry and 14% in populations of Asian ancestry.[14] To put it another way, over 90% of our human genomic data collected thus far is coming from only ~30% of the world's population (Fig. 4.1). Although the total amount of genomic data produced continues to increase, the proportion of that data coming from diverse populations outside of Europe and Asia has remained largely unchanged in the past 6 years.[16] This is a big issue for equity in both precision health and population genetic research as the structure of linkage disequilibrium varies by population, impacting our understanding of genetic risk factors and population history models. For example, recent work has demonstrated that PRS are currently far more accurate in European populations and do not port well across genetic ancestries.[17] Recent work including more diverse populations from developing countries has also helped refine our understanding of human origins. Ragsdale and colleagues used whole genome sequencing data from 12 African populations to infer detailed demographic models of early diversification in human populations, finding evidence for gene flow between two or more weakly differentiated ancestral *Homo* populations over hundreds of thousands of years before the earliest divergence in contemporary human lineages, some 120—135 thousand years ago.[18]

If a more complete picture of human evolution, present-day phenotypic and genomic diversity, and precision medicine are going to be realized, a more representative sample of human genetic variation must be included in the literature. However, collecting the genetic samples needed to achieve this diversity is easier said than done, as sample collection in developing countries can be much more challenging than in developed countries for a variety of ethical and logistic reasons. We use the developed/developing country nomenclature in this chapter but acknowledge the heterogeneity of nations within these categories and the potential problems with any international classification system

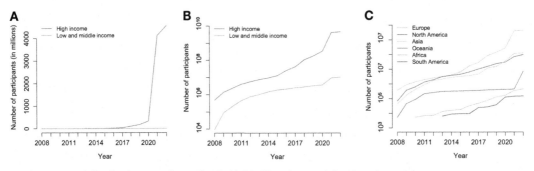

FIGURE 4.1 Participation in genomic studies is highly biased toward developed countries.

Cumulative number of participants in genome-wide association studies (A) on a linear scale and (B) on a logarithmic scale. Values are summed across all high versus low- and middle-income countries, as defined by the World Bank. (C) Cumulative number of participants in genome-wide association studies, on a logarithmic scale, summed across countries on each continent.

Data were downloaded from https://gwasdiversitymonitor.com/. Mills MC, Rahal C. The GWAS diversity monitor tracks diversity by disease in real time. Nat Genet. 2020;52(3):242–243. https://doi.org/10.1038/s41588-020-0580-y.

based on historical, financial, or developmental metrics.[19] On the logistical side, procurement of necessary supplies can be difficult and delayed, sampling must often be done outside of a centralized location, and transport of time-sensitive materials to laboratory facilities may take hours or even days longer.

Beyond the logistical challenges to actually conducting the research, those interested in increasing the diversity of genomic datasets must also grapple with the history of exploitative research and neo-colonial science that has taken place over the last century. Neo-colonial science,[20] commonly referred to as "helicopter science," refers to a practice whereby researchers from developed countries conduct research in developing countries without establishing collaborations with local researchers or acknowledging local collaborators who may have been involved in the data collection. In research involving human populations, this idea extends to research that does not establish meaningful connections with participant communities that allow them to help shape the research questions, derive tangible benefits when commercial products or therapeutics are developed, or engage with the research results beyond publications written for a scientific audience. Although this practice has been commonplace in all areas of science for well over a century, there is a growing recognition that this exploitative form of research has no place in 21st-century science.[21,22]

In human genomics, helicopter research has taken the form of researchers coming to a study site having only gained official permission (a permit often obtained from a national research body, public health authority, or institutional review board [IRB]), but not consulting the study population about the research. Once samples are collected, the researchers leave without working with local collaborators to extract, sequence, and analyze DNA samples. When the research is complete, the local research assistants and interpreters hired to assist in sample collection are not acknowledged for their contribution to the work. Local researchers working in this area may hear of the work for the first time at a conference or when a manuscript is published. The study participants and communities themselves are never contacted about the results of research and may only see media coverage of the research findings. One high-profile example of this comes from Brazil, where Indigenous groups such as the Karitiana and Surui donated blood to researchers decades ago, only to find out that cell lines had been derived from these samples and were being used in numerous research projects of which the communities were unaware.[23]

Despite continued calls for action and policy changes, the risk of helicopter science persists (but see,[24,25] for examples of recent structural progress). For those looking to start a new research project, navigating the process of forging mutually beneficial collaborations with local researchers and meaningful engagement with participant communities can be a daunting task. We argue that to establish a genomics project in developing countries, it is essential to forge mutually beneficial collaborations with local researchers and participant communities; this takes time and is a necessarily slow process. This reality stands in opposition to (1) the expectations of contemporary research, where the "publish or perish" mindset continues to be the norm as well as (2) the realities of grant budgets and timelines, which do not typically allow for the slow development of relationships.

In this chapter, we summarize best practices as well as many of the uncertain and challenging aspects of establishing a genomic project in developing countries; we also provide suggestions and examples of solutions to both common and emerging challenges. While our focus is on developing countries in particular, we note that many of the challenges and solutions discussed here are equally applicable to working with Indigenous and other marginalized or underrepresented groups in developed countries as well. In fact, much of the literature we draw from here has been driven by Indigenous peoples in the US (however, we note that a major issue is that Indigenous communities outside of the US often have even less access to political power and higher education,[26-28] and thus less ethics research has involved them or addressed their unique issues). We share our own experiences starting and maintaining long-term research collaborations with researchers and communities in developing countries and provide some helpful questions for one to ask at the beginning of a research project (Table 4.1). Our goals in doing so are to encourage other researchers to begin their own projects to diversify genetics research and to raise awareness to institutions and funders in developed countries about the unique challenges and lack of relevant structural support for implementing ethical, long-term research in developing countries.[29]

Table 4.1 Points to consider when engaging communities for genomics research.

Category	Example questions
Project goals and research questions	Does the community agree with the proposed research and overall project goals? Does the project include any specific questions that are of interest to community members? What benefits would the community derive from participating in the proposed research?
Consent	Is individual consent sufficient or should community consent also be obtained? How broadly should consent be applied? Do researchers have consent for future analyses, such as revisiting the same samples using different methods or technologies? In which future use cases should researchers seek to re-consent community members?
Methods	What questions do community members have about the methods used? How will sampling be conducted? Is the sampling process destructive, and if so, what does that entail?
What happens to samples during the project	Where will samples be stored? Who will have access to the samples? How will samples be processed?
What happens to samples after the project	Are any leftover products generated with the proposed research? Will leftover samples or products be stored, destroyed, or returned/repatriated?
Engagement and feedback	How often should researchers contact the community throughout the project period? Do community members want to receive regular updates and through what mechanism? Is there a plan to return results at project end?
Data dissemination, ownership, and privacy	How should the genetic data generated in this study be disseminated at project end? What safeguards are in place to ensure community privacy? Does the community consent to genomic data being posted in open databases or repositories? Do community members prefer to have restrictions on data sharing put in place? If so, what type of restrictions? Do community members want to be included in the administration of these data after project ends?

Community engagement in human genomics research: Where we've been and where we're going

Community engagement is the process of working equitably and collaboratively with study participants toward a shared goal or common interest.[30] Meaningful and sincere community engagement is essential for ethical human genomics research conducted anywhere in the world, especially in developing countries and/or when marginalized or underrepresented groups are involved. This is because many underrepresented populations have not only historically experienced a general lack of transparency surrounding genetics research, but even intentional harms and abuses in some cases. For example, in the 1990s, the Havasupai tribe participated in a research project with scientists at Arizona State University (ASU) to understand genetic risk for diabetes, which was devastating to the tribe and was thus of great community concern. However, the Havasupai later learned that their DNA had been used to study a wide range of topics they did not consent to, such as mental illness and demographic history, which conflicted with traditional narratives. The Havasupai took legal action against ASU and the university eventually settled with the tribe. This settlement was significant because it formally established what should have been already obvious—that study participants must be fully informed about how their DNA (or other data types) will be used in research, and that study communities should be meaningfully consulted about study design, project goals, and research outcomes.

The Havasupai story is unfortunately not unique (see[31−34]), and concerns about informed consent and a lack of community engagement in studies of underrepresented groups have permeated the fields of medical, evolutionary, and anthropological genetics for quite some time.[35] One recent example comes from a 2018 genomic study of the Bajau people in Indonesia. The publication of the study was met with skepticism from some Indonesian officials and researchers who argued that although national research permits were obtained, local permissions and proper engagement with communities were lacking.[33] These concerns have rightfully reduced enthusiasm for participating in genetic studies in many communities. In some cases, this history has led to blanket moratoriums on genetics research (e.g., the Navajo Nation[36]) or to specific codes of ethics that must be adhered to for any research to be considered (e.g., the San[37]). To move forward in our collective goal to increase diversity in genetics research, we must acknowledge past harms and address them head-on. Reassuringly, essentially all genetic subfields have experienced a growing sense of awareness and a call to action to reform community-engagement practices. For example, the American Society of Human Genetics recently published guidelines for working with underrepresented communities in ways that will "minimize harms and maximize benefits."[24] Similarly, the American Journal of Biological Anthropology requires that all genetic studies address a series of questions about community engagement and informed consent during the peer review process. Professional organizations in developing countries have also issued guidelines relevant to genetic research in these areas, such as the Biological Anthropology Association of Argentina's statement[38] on the ethical treatment of human remains and the International Association of Caribbean Archaeologists[39] best practices guide regarding destructive sampling and community engagement. Community-focused NGOs, such as the South African San Institute, have also issued guidelines for genetic sampling and community engagement.[40] More generally, there has been a flurry of academic writing on the topic of ethical research and community engagement in the last few years across all genetic subdisciplines.[21,41−43] While there is still significant work to be done, our personal view is that there is currently a clear sense of urgency and motivation on this topic among the genetics and genomics community at large.

So what does ethical (i.e., anti-helicopter/parachute) community engagement look like? Recent work by a diverse group of scientists has outlined best practices for genetics research in developing countries and/or that involves underrepresented communities.[21,24,30,44−46] For example, Claw et al. (2018) suggest the following best practice principles:

(1) Understand community sovereignty and research regulations
(2) Engage and collaborate
(3) Build cultural competencies
(4) Improve transparency of research practices
(5) Build local research capacity
(6) Disseminate research in accessible formats

Similarly, Lemke et al. (2022) outline the following guidelines for respectful, community-based research projects: before the work,

(1) Learn about the community and culture
(2) Identify and consult with community partners
(3) Codevelop research plans with stakeholders (including plans for communication, community benefits, and results return)
(4) During the work, seek out engagement and critical feedback from community members
(5) After the work, interpret and return results to the community
(6) Provide a mechanism to evaluate the community-engagement effort

The common thread behind these principles is the importance of building long-term, respectful, and mutually beneficial relationships. Another common thread is that communities are a dynamic and integral part of the entire research process. They are in essence a project collaborator, rather than a passive source of data or genetic material. In the same way that you would consult with a project collaborator about study design, any major protocol changes, and your ongoing results, communities should also be part of ongoing conversations about the direction, implementation, and interpretation of scientific results.

Common challenges for community-engaged genomics research

We are fully supportive and committed to the vision and best practices discussed above, but we acknowledge that implementing it does not come without challenges. Such challenges should not be an excuse for researchers to ignore certain principles, but we would like to be honest about potential hurdles in the interests of finding paths forward. One potential challenge for community-engaged research is that it can often be difficult to define and thus connect with "the community." In some cases, a given group of interest has a formal political or social structure that makes contacting and discussing research proposals with group members relatively obvious (e.g., tribal governments for Indigenous groups in the US). There may also be community-led organizations, such as NGOs or heritage groups, that can enable interfacing with a representative portion of the study community. However, in other cases, there may be no formal body or structure through which researchers can easily engage with an entire group in a systematic way. In these cases, conversations will need to be had on a location-by-location basis (e.g., at each household or village). This can create a situation in which there is no clear consensus among all study participants about certain aspects of the research plan; the best that researchers can do is to have as many conversations as possible, to acknowledge differing opinions, and to reiterate that anyone not in agreement with the current direction can opt out at any time. In our experience, this scenario is typically closer to reality, whether there is or is not a clear way to define "the community," as communities are always made up of individuals with varied viewpoints. This fact is helpful to acknowledge and anticipate when planning for a new community-based project. Our own work in Peru and Belize has been possible because of relationships with community-led NGOs who share common goals with the research projects.

A second major challenge for the vision laid out in Claw et al. (2018), Lemke et al. (2022), and elsewhere is that IRB requirements demand a minimal ethical standard, and thus complying with these oversight bodies does not ensure community-engaged research. Individual consent procedures and documents developed as part of the IRB process are not always tailored to local cultural contexts, and in essentially all cases more explanation is needed. For example, in our experience, community meetings at multiple levels are important for generating as much dialog as possible (e.g., meetings with all community members, households, or respected members that are appropriate to approach first such as elders or community governing bodies). Importantly, there are currently few incentives—institutional, funding, or publishing—to encourage researchers to go beyond the basic expectations of IRB. Structural changes will need to occur before we can reasonably expect universal compliance to the community-engagement principles discussed here.

Related to the above point, it is important for researchers starting new projects in developing countries to understand that most funding agencies do not explicitly support the community-engagement process. Building long-lasting and trustful partnerships requires a great deal of face-to-face time in the community prior to any data collection (e.g., to build cultural competencies and codevelop research plans). This investment is critical but is at odds with grant timelines, which are typically 2–5 years. Further, results return is an essential part of community-engaged research, but it can be more difficult to get support for this from funding agencies relative to support; for example, for laboratory reagents or genomic sequencing. Reassuringly, the National Science Foundation in the United States has approved travel for results return as part of recent research grants, and the Wenner-Gren Foundation has launched a specific funding mechanism for community-engaged partnerships and codeveloped projects in anthropology. We hope these examples signal that funding agencies are moving in the right direction, but until greater progress is made, it can be difficult for academic researchers to fund activities that are essential to community partnerships.

Finally, researchers must be prepared for their scientific goals to not be of interest to the target community, or for them to be of lower priority than other acute and immediate issues. Genomics research in developing countries can lead to crucial discoveries, but the benefits of this work that are relevant to study participants will likely take years to decades to materialize. In most cases, the benefits of genomics research will also manifest at the structural level (e.g., informing health policy or medical practices) rather than at the community or individual level. It is important to be honest about this reality throughout the research process; nevertheless, it can also be important to think about and include community benefits that extend beyond the main foci of the research itself (e.g., pairing health care or educational opportunities with genomic samples or data collection). The needs and desires of local communities will of course vary widely, but populations in developing countries often face intertwined challenges derived from systemic causes; for example, environmental degradation, threatened land rights, minimal access to education, health care, or other basic services, and limited economic opportunities. Rightfully so, the priorities of communities will thus seldom match perfectly with the scientific priorities of researchers. Working with communities to acknowledge and find solutions to their priorities, whether as part of the scientific goals of the study or not, is one clear way to build trust and create mutually beneficial relationships.

Working with local researchers in developing countries: Getting started

In addition to engaging with local communities, new genomics projects driven by researchers in developed countries, but focusing on communities in developing countries, must form equitable collaborations with local researchers as well as work toward building the research infrastructure in developing countries to make them independent. We discuss key themes along these lines, focusing on the perspective of someone from a developed country designing their first research project in a developing country. There are a few things in particular to which one needs to adjust, all of which local researchers are well-versed in and can lead the way via collaboration. Below, we focus on three key aspects: *different cultural norms, different available infrastructure, and different types of research and regulation practices.*

New researchers in a region need to take the time to understand the local history and the social context of the community they are working with. This is essential to make sure that you are sensitive to the potential impacts of the work you are doing. Many communities have been historically marginalized and targeted by discriminatory practices. We must be very careful that our work does not negatively impact them. Obtaining informed consent for genomics research can be complex in settings where there are lower literacy levels. The bioethical framework of individual consent may also be complicated as decision making in some communities is more hierarchical and group focused. Furthermore, some diseases that may be of interest to researchers, such as mental illness, are subject to greater stigma in some communities, requiring cultural awareness and sensitivity to differences.

Infrastructural problems can be a major impediment to starting and implementing research projects in developing countries relative to developed countries. These can include unreliable or no electricity in clinics and laboratories where samples will be collected and processed, areas that are difficult to access because of poor road conditions, and political instability making some areas inaccessible to researchers. Many countries do not have sufficient laboratory equipment or facilities for genomics research, particularly new and emerging technologies such as single-cell sequencing. Most developing countries must also import reagents to conduct genomics research, which can result in the work being much more expensive and time consuming than it would be in a developed country.[47,48] Access to high-speed internet and high-performance computing facilities is also limited in many developing countries, hampering bioinformatics training and analysis. High staff turnover can also be a challenge due to inadequate pay and competing demands for time on what are often too few qualified staff. Highly trained researchers from developing countries often feel pressure to leave to find better opportunities, making brain drain a major issue.

Obtaining the appropriate ethical approvals to conduct genomic sampling in developing countries comes with additional challenges. Doing this work in developed countries requires ethical approval from a single IRB (or several if more than one research institution is involved). Conducting similar studies in developing countries requires additional levels of approval before sample collection can begin. In our work in various countries in Africa, Latin America, and Southeast Asia, a local IRB is always required in addition to an IRB from one's home institution. Researchers often find themselves in a "catch-22," where both their home IRB and the local IRB want the other to be approved first before their final approval. This can take many weeks or months to resolve. This is another area where having a good relationship with local collaborators can be essential, as they are likely very familiar with the approval process, which may be confusing to those not working in the country. In addition to a local IRB, country or province-specific permits are often required to conduct research. For example, in Kenya, each PI needs individual-level research permits as well as a county-level permit to conduct research in a particular region. Depending on the country, these permits can take weeks or months to acquire, as the process is not always well-defined and review boards may meet infrequently. In some countries, permits need to be renewed yearly, and for interdisciplinary projects (for example, studying both population history and biomedical questions), researchers may need permission from multiple government bodies such as the Ministry of Culture and the Ministry of Public Health. Processes for submission and renewal of permits can be unclear to outside researchers and could take much longer to obtain than in developed countries.

Ultimately, overcoming all of these issues requires an enormous amount of patience. One cannot expect the research to go along the same timelines as research done in developed countries, so it is important to build safeguards for time into the research schedule. One good rule of thumb is to take the amount of time you think it will take to complete the project and triple it to get a more realistic timeline. Plan to go to the host county for large amounts of time to build relationships before starting to collect any data. This is part of the process and an essential foundation for everything else. Having a trusted local collaborator is invaluable to navigating the ethical approval process and in making sure that the research questions and published results are sensitive to historical and social issues that might surround a particular topic or community. For example, ethnic nomenclature is a subject of socio-political debate in South Africa. Tens of thousands of South Africans today self-identify as members of a Khoe or San group. The historical term used by anthropologists to describe these communities is "Bushman," but given the obvious derogatory implications of the term, many anthropologists and population geneticists today use "Khoe-San" to refer to the San and Khoe groups collectively. However, the San Council of Southern Africa prefers to keep these terms separate (i.e., San and Khoe) to denote the different cultures. Many "San" individuals prefer being called "Bushmen," while others consider the word to be pejorative. Many other individuals with recent Khoe and San ancestors self-identify as "Colored," a South African ethnic category that can trace its history to the colonial and Apartheid periods. While complete consensus on the proper terms to use for a particular community is unlikely, having a local collaborator, tuned into the ongoing discourse is essential in navigating this complex topic. The value that this local expertise and perspective provide on a research project cannot be understated. However, it is important to engage with researchers in developing countries in a mutually beneficial way, not just to enlist them as guides through common hurdles.

Equitable collaborations with researchers in developing countries

There is a growing recognition among scientists that equitable collaboration should be a priority for all research projects, especially those between scientists in developed countries and developing countries. Too often, collaborators from developing countries in international projects are treated as solely data collectors or a means to local permits and permissions.[49] Their research questions and funding needs are not considered when writing papers and applying for grants. Many solutions to this helicopter research that have been used in the past have been done without a long-term perspective of what benefits they would actually provide local research teams. For example, donating or otherwise securing lab equipment can bridge the gap in resources between researchers in developed and developing countries, but without dependable electricity, or continued financial support to train and hire technicians, perform routine maintenance, and purchase expensive reagents, the equipment is not likely to be used after the end of the project.

Researchers starting new collaborations with scientists in developing countries must intentionally work to build engagement and equity into every aspect of the partnership by maintaining regular communication and evaluating behavior. Starting a project is a long-term commitment, so it can be daunting for young researchers, who were likely trained on established projects, to take the first steps. There will always be some trial and error in this process, so maintaining continuous feedback between all collaborators on a project is critical for solidifying relationships, and identifying areas for improvement. The partnerships needed to conduct a successful and equitable research project require trust that can take years to build and may not be easily repaired once broken. Candid conversations and regular check-ins between collaborators will help to ensure that minor setbacks do not turn into project-ending obstacles. We provide three aspects of equitable collaborations that we see as vital to their success. We emphasize that the particulars of any project will vary by the particular context, so do not offer a one-size-fits-all strategy.

(1) Shared decision making and leadership with regard to all aspects of a research project, including establishing research questions, data collection, and analysis. This also means meaningful involvement in funding opportunities for all collaborators. Including international collaborators in grant applications can be very painful administratively, but it is important that collaborators in developing countries have real budgets to fund shared work.
(2) Continual engagement throughout the research process, not just asking for samples and then only recontacting collaborators at the time of manuscript drafting. Instead, aim to have regular meetings for keeping each other abreast of project progress and difficulties.
(3) Creating a team environment where differing perspectives are respected and considered. Sometimes compromises need to be made to prioritize goals of all researchers and make sure everyone's research program is well-represented in the final work. Creating this environment requires honesty and trust between collaborators and a willingness and flexibility to accommodate each other's needs.

It can be useful to establish detailed collaboration agreements at the outset of the project that lay out explicit responsibilities for each collaborator and lay ground rules such as who will be leading particular parts of the project when and how data will be made accessible to other researchers, and authorship policies.

Study designs for genomic research in developing countries

As laid out earlier in the chapter, working in developing countries poses unique challenges to genetic sample collection that need to be taken into account during study design. The majority of population and epidemiological genetics research studies that involve sample collection are designed as observational studies. These fall into three categories: cross-sectional studies, case-control studies, and longitudinal cohort studies. Cross-sectional studies, by looking at a population at a single time point, are cost-effective and allow genetic epidemiologists to look at the genetic variation associated with a particular phenotype or the prevalence of a disease of interest. Population genetic studies are also usually cross-sectional, using the genetic data collected from unrelated individuals at a single time point to help reconstruct the genetic history of a population. Along the same lines, case-control studies also enroll participants at a single time point but focus on participants with a single condition and matched controls without the condition. Over the past 10 years, many countries and universities have established large biobanks to link participant genomic data with biomarkers and often electronic health records to more easily untangle the complex relationship between genetics and disease[50] using these two study designs. Many developing countries have not yet switched to centralized electronic medical records and in many cases do not have the resources and infrastructure to enable this, producing a lag behind efforts in developed countries. One great exception to this rule is the General Population Cohort in Uganda.[51,52] Originally established in 1989, and moved to an electronic records system in 2010, this cohort has led to the production of the Uganda Genome Resource, genome-wide data on over 5000 individuals linked with demographic, biomarker, and clinical phenotype data.

Longitudinal studies are ideal for studying causal relationships between many social and environmental factors and diseases of interest. While establishing and maintaining such a study is challenging anywhere, conducting longitudinal research in developing nations provides some unique challenges. Recontacting participants to collect new data throughout the study duration can be difficult because in some areas people tend to move around more either within a community or between communities for work or due to nomadic subsistence strategies. In our projects in rural South Africa, we have a recontact rate of <50%, primarily because people move to larger towns for work. In our work in rural Kenya with Turkana pastoralists, the recontact rate is even lower because traditional subsistence practices are based on a nomadic lifestyle. Unlike in developed countries, researchers cannot take for granted that participants will always have a stable email address and phone number, further complicating recontact efforts.

Despite the numerous challenges to conducting research in developing countries, some study designs may actually be easier to deploy in these countries. In South Africa, one of the authors and their collaborators have been working on population history projects with Khoe-San and Khoe-San descendant communities for well over a decade. After obtaining the appropriate ethical approvals, potential participant communities are contacted up to a year in advance of sampling to introduce them to the project. Community meetings are scheduled to get feedback from community members about any interests and concerns they have. After a short period of time, we return to the town, and with the help of trained local research assistants, visit the homes of community members to enroll them in the study. Over 95% of people approached in this "house-to-house" model have enrolled in the study over the years. This strategy works best in small communities with strong social bonds and would likely not port well to many settings in developed countries where there is less community and social connection among people who live in a given place.

New researchers who become involved in the South Africa project described above, who are used to working in developed countries, are often surprised at the willingness of people to let the research team into their homes and participate in the study. There are likely several reasons for this, the discussion of which is beyond the scope of this chapter. However, we have recently learned through a disease-centered study in the same region, that a house-to-house sampling strategy is familiar to many community members already, perhaps paving the way for our own sample collection. Health clinics in South Africa employ community health workers who regularly go into the community to sample household contacts of tuberculosis patients (collecting information and sputum samples), among other duties. It is common for patients to provide medical and demographic information, as well as biological (typically sputum) samples to the community healthcare workers. This familiarity with healthcare workers visiting people's households makes the collection of random community sampling much easier than it would be in developed countries. Multiple generations of a family and occasionally friends often live in the same household, making the collection of family-based cohorts for genetic studies much easier in South Africa (and other developing countries) than in developed countries.

Ongoing and emerging challenges to genomics research in developing countries
Working during pandemics

The COVID-19 pandemic posed significant challenges for field-based scientific research over the last several years.[53–56] Beyond its impact on researchers, communities in developing countries experienced unequal access to preventive care and healthcare services, increased economic precarity, and heightened vulnerability to the compounded risks posed by the health crisis.[57–59] In this context, some have called for reassessing the role and need for fieldwork in scientific research and prompted field-based researchers to consider how to modify projects in ways that are sensitive to the needs and challenges of participant communities.[60,61] This call to action is especially pertinent at the present moment as researchers from Global North nations, where vaccines and therapeutics for COVID-19 are now widely available, begin returning to the field and coming into contact with communities where access to these measures may still be limited and vulnerability to infection remains high.[62]

In the authors' own experiences, the COVID-19 pandemic caused significant disruption to both ongoing and new research projects. These included halting data collection (in some cases for well over a year); suspending travel to field sites; diverting research funds toward unexpected expenses, such as purchasing personal protective equipment or assisting local collaborators, and relying on internet or phone communication to meet with research participants. The pandemic also prompted our teams to innovate alternative approaches for continuing research in the absence of in-person interactions. One of the authors found that remote meetings provided an accessible and low-cost means to remain connected to participant communities for consultation, regular project updates, and returning results. The virtual format also allowed for increased transparency because the research team was able to deploy visual media, such as animations and video, to demonstrate ongoing laboratory processes. We note, however, that these approaches are heavily dependent on reliable internet access and bandwidth—which may not be available in all locations[63,64]—and should be considered a complement, not a replacement, for in-person meetings with participants.

Genomic data access and secondary use

A growing culture of transparency and data reuse has moved most fields, including genomics, toward the norm of making all datasets public (usually following publication). Open data allows the field to evaluate reproducibility and also enables researchers to ask new questions via aggregated datasets. More generally, the open science movement believes that open data will lead to better research, better accountability and connection to non-scientists, and ultimately greater societal benefits[65]; in fact, work funded by the National Institutes of Health in the United States and the European Research Council is now required to post all underlying data (with a few exceptions) to open access platforms.[66]

While the potential benefits of open science are indeed plentiful, making genomic data generated from human populations in developing countries publicly available is a complicated issue because there is great potential for inappropriate or unfair use of the data. For example, the data could be used for purposes that do not benefit study communities or align with their priorities (e.g., the Havasupai case discussed earlier in the chapter), uses that lead to identification of anonymous subjects or groups in small populations, and uses with potentially stigmatizing interpretations. While the field of genomics is and should be committed to open and reproducible science, above all else it should be committed to protecting study communities, prioritizing their interests, and empowering their stewardship over their own data. We therefore encourage genomics researchers working in developing countries, or with any Indigenous, underrepresented, or otherwise marginalized groups, to discuss issues of data sovereignty, data access, and secondary data use during study design and informed consent. As a set of guiding principles to start these discussions, we refer readers to the CARE Principles for Indigenous Data Governance, which broadly encompass the ideas of (1) collective benefit, (2) authority to control, (3) responsibility, and (4) ethics.[67]

Given the issues and sensitivities discussed above, in most cases, the appropriate path will be to not make the data public without restriction and instead to only allow access to researchers who agree to uphold the original IRB and community agreements.[68–70] However, current issues with implementing this model are that (1) it is often not formalized, and the procedures for obtaining access typically involve emailing the authors of a given study and (2) it is often not clear how decisions are made about data reuse and how communities themselves are involved. We are therefore encouraged by recent movement toward a more formalized controlled access model, in which researchers interested in secondary data use apply and are reviewed by a board composed of primary stakeholders (e.g., principal investigators, local scientists, and community members, see discussions in Claw et al., Hudson et al., Collier-Robinson et al., and Mackey et al.[21,46,71,72] for an example). For example, to access the genomic data produced as part of the Indonesian Genomic Diversity Project, researchers must submit an application to a data access committee led by Indonesian researchers. As with all aspects of genomics research in developing countries, one size will never fits all; given the push toward open data, the tensions this can create with the priorities and protection of study communities and the scant number of established models or successful examples, we encourage researchers to make this area a priority for discussion during study design.

Building genomics research capacity in developing countries

An important consideration for any collaboration between researchers in developing countries and developed countries is access to data, infrastructure, and support so that publication is also equitable. Too often genomics papers are led by researchers primarily from developed countries who have better wet-lab, computational, administrative, or other types of support to produce these papers quickly. Part of this disparity comes from a divide in data analysis expertise and long-term support for building local institutional knowledge. A common solution to bridge this gap with collaborators in developed countries is to run intensive short courses, roughly week-long classes on skills associated with the research. Although these workshops are a useful way to bring people together and transfer skills, if the topics taught are not immediately applicable to the data produced by the project, the courses risk not having a lasting effect. A more sustainable model, perhaps, is codesigning short courses with local faculty to "train the trainers" and produce training materials geared toward local needs and capabilities with long-term usage in mind. Those establishing data-sharing policies for outside researchers may consider a longer embargo period to enable all collaborators to analyze data and publish findings. Establishing these parameters from the beginning of a project as part of a collaboration agreement can help negate unnecessary tension as a project progresses.

The NeuroGAP-Psychosis research project, a large genomic consortium linking dozens of researchers in four African countries with collaborators in the US and UK, recently published a summary of their own training program, GINGER.[73] As part of this program, all principal investigators designed a hybrid training program to train and support a team of local leaders in mental health genetics. GINGER consists of three components to train new PIs, undergraduate and graduate students, and produce institutional curriculum that can be used in existing courses at the African training institutions. Another example is the Capacity Building for Bioinformatics in Latin America (CABANA) project, which ran from 2017 to 2022. As the name suggests, this joint initiative between nine universities across Latin America and the European Bioinformatics Institute aimed to build bioinformatics capacity in the region with a focus on research themes determined by the Latin American partner institutions.[74] While these programs were conducted on a scale larger than many projects require, they can serve as a model for future programs to design useful and sustainable training programs at an appropriate scale.

Conclusions

While there are several challenges to conducting genomic research in developing countries, we believe there are also many unique opportunities: to better understand the genetic architecture of human biology and disease; to uncover genetic adaptation to diverse environments; to recreate historic patterns of colonization, migration, and demography; and to address the impacts of changing lifestyles on health. Achieving these scientific goals will also address the need for increased diversity and representation in both participant and researcher communities. Thus overcoming the challenges of implementing research in developing countries warrants the investment of time and resources. Sample collection or fieldwork in developing countries is often thought of as a special class of research work, but what we have discussed in this chapter should be the norm for genomics projects and is widely applicable to research collaborators anywhere with Indigenous, underrepresented, or otherwise marginalized communities.

In this chapter, we have emphasized two broad areas of particular importance to increasing the diversity of genomic datasets in an ethical way. First, genomic studies should meaningfully consult with participant communities about study design, project goals, and research outcomes before the start of sample collection. Researchers should be open to expanding the scope of a study to include participant input on these goals. Researchers must also return results to communities as the research progresses and discuss any potential benefits that may come at the end of a study. Second, it is essential to build mutually beneficial collaborations with local researchers, where the research questions and funding needs of both parties are given importance, and the research infrastructure, both physical and intellectual, can be expanded and shared. Our experiences and suggestions are not meant to be a universal framework—instead, we argue that the particular approach that should be used will be project- and community-dependent. Only through consultation and relationship-building between researchers and participant communities can we hope to create more equitable outcomes from genomic research.

References

1. García-Ortiz H, Barajas-Olmos F, Contreras-Cubas C, et al. The genomic landscape of Mexican Indigenous populations brings insights into the peopling of the Americas. *Nat Commun.* 2021;12(1):5942.
2. Gopalan S, Berl REW, Myrick JW, et al. *Curr Biol.* 2022;32(8). https://doi.org/10.1016/j.cub.2022.02.050, 1852-1860.e5.
3. Nieves-Colón MA, Pestle WJ, Reynolds AW, et al. *Mol Biol Evol.* 2020;37(3):611−626. https://doi.org/10.1093/molbev/msz267.
4. Green RE, Krause J, Briggs AW, et al. *Science.* 2010;328(5979):710−722. https://doi.org/10.1126/science.1188021.
5. Sankararaman S, Mallick S, Dannemann M, et al. The genomic landscape of Neanderthal ancestry in present-day humans. *Nature.* 2014;507(7492):354−357.
6. Meyer M, Kircher M, Gansauge M-T, et al. *Science.* 2012;338(6104):222−226. https://doi.org/10.1126/science.1224344.
7. Lea AJ, Akinyi MY, Nyakundi R, et al. *Proc Natl Acad Sci U S A.* 2018;115(52):E12163−E12171. https://doi.org/10.1073/pnas.1811967115.
8. Lea AJ, Waigwa C, Muhoya B, et al. *Evolut Med Public Health.* 2021;9(1):406−419. https://doi.org/10.1093/emph/eoab039.
9. Wallace IJ, Lea AJ, Lim YAL, et al. *BMJ Open.* 2022;12(9). https://doi.org/10.1136/bmjopen-2021-058660. e058660.
10. Lewis CM, Vassos E. Polygenic risk scores: from research tools to clinical instruments. *Genome Med.* 2020; 12(1):44.
11. Khera AV, Chaffin M, Aragam KG, et al. Genome-wide polygenic scores for common diseases identify individuals with risk equivalent to monogenic mutations. *Nat Genet.* 2018;50(9):1219−1224.
12. Schumacher FR, Al Olama AA, Berndt SI, et al. Association analyses of more than 140,000 men identify 63 new prostate cancer susceptibility loci. *Nat Genet.* 2018;50(7):928−936.
13. Sharp SA, Rich SS, Wood AR, et al. *Diab Care.* 2019;42(2):200−207. https://doi.org/10.2337/dc18-1785.
14. Popejoy AB, Fullerton SM. Genomics is failing on diversity. *Nature.* 2016;538(7624):161−164.
15. Mills MC, Rahal C. The GWAS Diversity Monitor tracks diversity by disease in real time. *Nat Genet.* 2020; 52(3):242−243.
16. Fatumo S, Chikowore T, Choudhury A, Ayub M, Martin AR, Kuchenbaecker K. A roadmap to increase diversity in genomic studies. *Nat Med.* 2022;28(2):243−250.
17. Martin AR, Kanai M, Kamatani Y, Okada Y, Neale BM, Daly MJ. Clinical use of current polygenic risk scores may exacerbate health disparities. *Nat Genet.* 2019;51(4):584−591.
18. Ragsdale AP, Weaver TD, Atkinson EG, et al. A weakly structured stem for human origins in Africa. *Nature.* 2023;617(7962):755−763.
19. Khan T, Abimbola S, Kyobutungi C, Pai M. *BMJ Glob Health.* 2022;7(6). e009704.
20. Dahdouh-Guebas F, Ahimbisibwe J, Van Moll R, Koedam N. Neo-colonial science by the most industrialised upon the least developed countries in peer-reviewed publishing. *Scientometrics.* 2003;56(3):329−343.
21. Claw KG, Anderson MZ, Begay RL, et al. A framework for enhancing ethical genomic research with Indigenous communities. *Nat Commun.* 2018;9(1):2957.

22. Nordling L. African scientists call for more control of their continent's genomic data. *Nat News.* 2018. https://doi.org/10.1038/d41586-018-04685-1.

23. Rohter L. *In the Amazon, Giving Blood but Getting Nothing.* The New York Times; 2007.

24. Lemke AA, Esplin ED, Goldenberg AJ, et al. *Am J Hum Genet.* 2022;109(9):1563−1571. https://doi.org/10.1016/j.ajhg.2022.08.005.

25. Nordling L. *PLOS Launches Policy to Weed Out 'helicopter Science.' Research Professional News*; 2021. https://researchprofessionalnews.com/rr-news-africa-pan-african-2021-9-plos-launches-policy-to-weed-out-helicopter-science/.

26. Fisher F. Improving education of indigenous people. In: *Issue Brief for the World Conference ODUMUNC 2015*; 2015. https://ww1.odu.edu/content/dam/odu/offices/mun/2014/unwc/wc-improving-education-of-indigenous-people.pdf.

27. *Indigenous Peoples' Right to Education.* United Nations International Day of the World's Indigenous Peoples; 2016. https://www.un.org/esa/socdev/unpfii/documents/2016/Docs-updates/Backgrounder_Indigenous_Day_2016.pdf.

28. Rice R. *Indigenous Political Representation in Latin America.* 2017. https://doi.org/10.1093/acrefore/9780190228637.013.243.

29. Precision medicine needs an equity agenda. *Nat Med.* 2021;27(5):737. https://doi.org/10.1038/s41591-021-01373-y.

30. Tindana P, de Vries J, Campbell M, et al. Community engagement strategies for genomic studies in Africa: a review of the literature. *BMC Med Ethics.* 2015;16(1):24. https://doi.org/10.1186/s12910-015-0014-z.

31. Dodson M, Williamson R. *J Med Ethics.* 1999;25(2):204. https://doi.org/10.1136/jme.25.2.204.

32. Greely HT. Human genome diversity: what about the other human genome project? *Nat Rev Genet.* 2001; 2(3):222−227. https://doi.org/10.1038/35056071.

33. Rochmyaningsih D. Did a study of Indonesian people who spend most of their days under water violate ethical rules? *Sci News*; 2018. https://www.science.org/content/article/did-study-indonesian-people-who-spend-their-days-under-water-violate-ethical-rules.

34. Rochmyaningsih D. *The Philippines Controversy and the Pitfalls of International Genomics Research.* The Wire; 2022. https://science.thewire.in/the-sciences/uppsala-study-philippines-population-genetics-pitfalls-international-genomics-research/.

35. Malhi RS. In: *A Companion to Anthropological Genetics.* 2019:37−44. https://doi.org/10.1002/9781118768853.ch3.

36. Claw KG, Dundas N, Parrish MS, et al. Perspectives on genetic research: results from a Survey of Navajo community members. *Front Genet.* 2021;12. https://www.frontiersin.org/articles/10.3389/fgene.2021.734529.

37. Nordling L. San people of Africa draft code of ethics for researchers. Science Insider; 2017. https://www.science.org/content/article/san-people-africa-draft-code-ethics-researchers.

38. *Declaración de la Asociación de Antropología Biológica Argentina (AABA) en Relación Con la Ética Del Estudio De Restos Humanos*; 2007. https://asociacionantropologiabiologicaargentina.org.ar/wp-content/uploads/sites/9/2019/12/Declaraci%C3%B3n-AABA-Restos-Humanos.pdf. Accessed July 18, 2023.

39. *International Association for Caribbean Archaeology (IACA) Code of Ethics*; 2022. https://bpb-us-e1.wpmucdn.com/blogs.uoregon.edu/dist/3/13484/files/2023/06/IACA-Code-of-Ethics-ENGLISH-FINAL.pdf. Accessed July 18, 2023.

40. South African San Institute. *Guidelines, Consent Instruments, Procedures and Protocols for DNA Sampling with San Traditional Communities in Namibia*; 2016. https://12ebe7cc-83ea-938a-ff56-b1dc3f7ac632.filesusr.com/ugd/ecae59_00d10f74086441d9ad62736cdd000667.pdf. Accessed July 18, 2023.

41. Borrell LN, Elhawary JR, Fuentes-Afflick E, et al. *N Engl J Med.* 2021;384(5):474−480. https://doi.org/10.1056/NEJMms2029562.

42. Kaladharan S, Vidgen ME, Pearson JV, et al. *BMJ Glob Health*. 2021;6(11). https://doi.org/10.1136/bmjgh-2021-007259. e007259.

43. Lewis KL, Turbitt E, Chan PA, et al. *Am J Hum Genet*. 2021;108(5):894−902. https://doi.org/10.1016/j.ajhg.2021.04.002.

44. Bankoff RJ, Perry GH. Genet Hum Origin. 41. 2016:1−7. https://doi.org/10.1016/j.gde.2016.06.015.

45. Garrison NA, Hudson M, Ballantyne LL, et al. Annu Rev Genom Hum Genet. 2019;20(1):495−517. https://doi.org/10.1146/annurev-genom-083118-015434.

46. Hudson M, Garrison NA, Sterling R, et al. Rights, interests and expectations: indigenous perspectives on unrestricted access to genomic data. *Nat Rev Genet*. 2020;21(6):377−384. https://doi.org/10.1038/s41576-020-0228-x.

47. Kalergis AM, Lacerda M, Rabinovich GA, Rosenstein Y. *Trends Mol Med*. 2016;22(9):743−745. https://doi.org/10.1016/j.molmed.2016.06.013.

48. Ávila-Arcos MC, de la Fuente Castro C, Nieves-Colón MA, Raghavan M. Recommendations for sustainable ancient DNA research in the global south: voices from a new generation of paleogenomicists. *Front Genet*. 2022;13. https://www.frontiersin.org/articles/10.3389/fgene.2022.880170.

49. Argüelles JM, Fuentes A, Yáñez B. *Am Anthropol*. 2022;124(1):130−140. https://doi.org/10.1111/aman.13692.

50. Wojcik GL. Cell. 2022;185(23):4256−4258. https://doi.org/10.1016/j.cell.2022.10.010.

51. Asiki G, Murphy G, Nakiyingi-Miiro J, et al. *Int J Epidemiol*. 2013;42(1):129−141. https://doi.org/10.1093/ije/dys234.

52. Fatumo S, Mugisha J, Soremekun OS, et al. Cell Genomics. 2022;2(11). https://doi.org/10.1016/j.xgen.2022.100209.

53. Forrester N. How to manage when your fieldwork is cancelled. *N*at News; 2020. https://www.nature.com/articles/d41586-020-03368-0.

54. Gao J, Yin Y, Myers KR, Lakhani KR, Wang D. Potentially long-lasting effects of the pandemic on scientists. *Nat Commun*. 2021;12(1):6188. https://doi.org/10.1038/s41467-021-26428-z.

55. Krause P, Szekely O, Bloom M, et al. COVID-19 and fieldwork: challenges and solutions. *PS Political Sci Polit*. 2021;54(2):264−269. https://doi.org/10.1017/S1049096520001754.

56. Pennisi E. How COVID-19 has transformed scientific fieldwork. Science Insider; 2021. https://www.science.org/content/article/how-covid-19-has-transformed-scientific-fieldwork.

57. Bayati M, Noroozi R, Ghanbari-Jahromi M, Jalali FS. Inequality in the distribution of Covid-19 vaccine: a systematic review. *Int J Equity Health*. 2022;21(1):122. https://doi.org/10.1186/s12939-022-01729-x.

58. Egger D, Miguel E, Warren SS, et al. *Sci Adv*. 7(6):eabe0997. https://doi.org/10.1126/sciadv.abe0997.

59. Phillips CA, Caldas A, Cleetus R, et al. Compound climate risks in the COVID-19 pandemic. *Nat Clim Change*. 2020;10(7):586−588. https://doi.org/10.1038/s41558-020-0804-2.

60. Kaplan HS, Trumble BC, Stieglitz J, et al. *Lancet*. 2020;395(10238):1727−1734. https://doi.org/10.1016/S0140-6736(20)31104-1.

61. Rudling A. Now is the time to reassess fieldwork-based research. *Nat Hum Behav*. 2021;5(8):967. https://doi.org/10.1038/s41562-021-01157-x.

62. Wood E, Rogers D, Sivaramakrishnan K, Almeling R. *Resuming Field Research in Pandemic Times*. Items: Insights from the Social Sciences; 2020. https://items.ssrc.org/covid-19-and-the-social-sciences/social-research-and-insecurity/resuming-field-research-in-pandemic-times/.

63. Broom D. *Coronavirus has Exposed the Digital Divide like Never Before*. World Economic Forum; 2020. https://www.weforum.org/agenda/2020/04/coronavirus-covid-19-pandemic-digital-divide-internet-data-broadband-mobbile/.

64. García-Escribano M. *Low Internet Access Driving Inequality. IMF Blog*; 2020. https://www.imf.org/en/Blogs/Articles/2020/06/29/low-internet-access-driving-inequality#:~:text=The%20lack%20of%20universal%20and,have%20more%20limited%20Internet%20access.

65. Amann RI, Baichoo S, Blencowe BJ, et al. *Science*. 2019;363(6425):350−352. https://doi.org/10.1126/science.aaw1280.

66. National Institutes of Health. *Final NIH Policy for Data Management and Sharing*; 2020. https://grants.nih.gov/grants/guide/notice-files/NOT-OD-21-013.html.

67. Carroll SR, Garba I, Figueroa-Rodríguez OL, et al. The CARE principles for indigenous data governance. *Data Sci J*. 2020;19(1):1−12. https://doi.org/10.5334/DSJ-2020-043.

68. Garcia AR, Finch C, Gatz M, et al. APOE4 is associated with elevated blood lipids and lower levels of innate immune biomarkers in a tropical Amerindian subsistence population. *Elife*. 2021;10:e68231. https://doi.org/10.7554/eLife.68231.

69. Keith MH, Flinn MV, Durbin HJ, et al. *PLoS One*. 2021;16(11). https://doi.org/10.1371/journal.pone.0258735. e0258735.

70. Reynolds AW, Mata-Míguez J, Miró-Herrans A, et al. *Proc Natl Acad Sci*. 2019;116(19):9312−9317. https://doi.org/10.1073/pnas.1819467116.

71. Collier-Robinson L, Rayne A, Rupene M, Thoms C, Steeves T. Embedding indigenous principles in genomic research of culturally significant species. *N Z J Ecol*. 2019;43(3):1−9. Accessed November 29, 2022.

72. Mackey TK, Calac AJ, Chenna Keshava BS, Yracheta J, Tsosie KS, Fox K. *Cell*. 2022;185(15):2626−2631. https://doi.org/10.1016/j.cell.2022.06.030.

73. Martin AR, Stroud RE, Abebe T, et al. Increasing diversity in genomics requires investment in equitable partnerships and capacity building. *Nat Genet*. 2022;54(6):740−745. https://doi.org/10.1038/s41588-022-01095-y.

74. Stroe O. Building bioinformatics capacity in Latin America. *EMBLetc*. 2022;99. https://www.embl.org/news/embletc/issue-99/building-bioinformatics-capacity-in-latin-america/.

Next-generation sequencing technologies: Implementation in developing countries

5

Nadia Carstens[1,3], Brigitte Glanzmann[2,3], Craig Kinnear[2,3] and Caitlin Uren[4]

[1]*Division of Human Genetics, National Health Laboratory Service and School of Pathology, Faculty of Health Sciences, University of the Witwatersrand, Johannesburg, South Africa;* [2]*South African Medical Research Council Centre for Tuberculosis Research, Division of Molecular Biology and Human Genetics, Faculty of Medicine and Health Sciences, Stellenbosch University, Cape Town, South Africa;* [3]*South African Medical Research Council, Genomics Platform, Cape Town, South Africa;* [4]*South African Medical Research Council Centre for Tuberculosis Research, Division of Molecular Biology and Human Genetics, Department of Biomedical Sciences, Stellenbosch University, Stellenbosch, South Africa*

Introduction

The Human Genome Project (HGP) was completed two decades ago and was hailed as a major scientific breakthrough comparable to the first moon landing.[1] Sequencing of the first human genome took 13 years to complete and involved the efforts of many scientists from several countries at a cost of approximately $300 million. The reason for this was that the only sequencing technology available at the time was Sanger sequencing and, while this technology is accurate and able to produce relatively long reads, it is expensive and slow and therefore its use for genome sequencing was limited to large highly funded sequencing centers. These centers were limited to developed countries.

Following the completion of the HGP, there was a concerted effort in the field of genome sequencing to develop more cost-effective sequencing technologies, which led to the era of next-generation sequencing (NGS).[2] NGS was a massive leap forward for sequencing technology as it made sequencing entire genomes possible in short periods of time at a reasonable cost. Moreover, it enabled scientists to expand on sequencing applications in several fields of research.

Despite all the advances in sequencing technologies since the completion of the HGP and the incredible reduction in the cost of NGS, this technology is still out of reach for many in developing countries. As a result, much of the focus to understand human disease has been on populations from high-income countries (HICs) with less than 2% of human genomes analyzed to date being from Africa despite the high genetic diversity of African populations.[3,4] Furthermore, many developing countries are burdened with the high prevalence of infectious disease, and as was demonstrated by the COVID-19 pandemic, genome surveillance plays a critical role in response to infection outbreaks. So, what are the reasons for the disparities in sequencing capability between developed and developing nations?

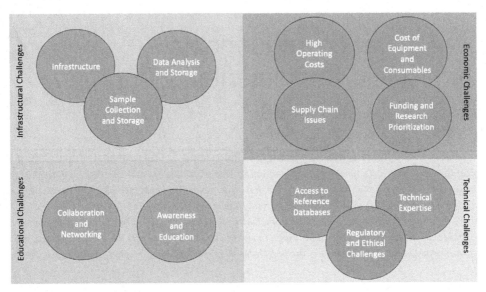

FIGURE 5.1 Barriers to sequencing.

Barriers can be broadly categorized into economic, infrastructural, technical, and educational challenges.

Created by Caitlin Uren.

Barriers to NGS access in developing countries

In their 2016 study, Heling and colleagues used data from the Genome Online Database (GOLD) and from the World Bank to evaluate the use of sequencing technology in developing countries.[5] GOLD is a comprehensive curation of genome sequencing projects worldwide.[6] By using data from GOLD coupled with the average research and development expenditure of each country, they showed that developed countries with more research and development expenditure had more sequencing centers and were therefore able to sequence more genomes. Additionally, developing countries with higher research and development expenditures were also more active in genomic sequencing compared to other developing countries that spent less on research and development. This suggests that limited funding is a major barrier to implementation of sequencing in developing countries.

A recent survey conducted to investigate the state of genomic medicine in Africa as well as a study investigating the impact of sequencing intensity and turn-around times on SARS-CoV-2 variant detection in 189 countries provided further insights into the barriers to sequencing technologies faced by researchers in Africa.[7,8] These barriers can broadly be categorized into economic, infrastructural, technical, and educational challenges, some of these key barriers are discussed below (Fig. 5.1).

Lack of infrastructure and high operating costs

Adequate infrastructure is crucial for NGS implementation, including a stable power supply, temperature-controlled environments, and access to high-speed internet for data analysis and sharing. Many developing countries lack these essential infrastructure elements.

Furthermore, one of the biggest hurdles for developing countries wanting to implement NGS is access to adequate laboratory infrastructure. In their survey, Jongeneel and colleagues noted that less than half of their responders had access to local infrastructure needed for genomic research and only 28% had access to local sequencing and computational facilities.[8] Moreover, infrastructure challenges are not limited to adequate laboratories, but also to computer infrastructure. NGS produces large amounts of data generated and the specialized skills, reliable broadband internet access, expensive computers, and data servers are not always available in many developing countries, particularly in Africa.[9]

Apart from the initial capital investment, NGS has ongoing operational costs related to maintenance. Maintenance costs are rarely included in grant funding and are largely overlooked by internal, institutional funding. A lack of local support could also mean operating at a loss due to a lack of samples to process.

High cost of equipment and consumables—From sample collection to data generation and storage

The cost of NGS implementation can be prohibitively expensive, particularly for developing countries with limited financial resources. These high costs not only apply to the data generation but also from the start of the project (sample collection) to the end (data analysis and storage).

Proper collection and storage of biological samples are crucial for efficient NGS implementation. Developing countries may lack the necessary facilities and expertize for proper sample collection, storage, and general handling. This can, in the end, lead to poor-quality data. Furthermore, the cost of sample collection in general is considerably higher in developing countries due to increased consumable costs and laborious field work.

Once the samples are collected, the DNA extraction process is followed by further laboratory procedures before sample loading on the sequencer. The majority of the equipment and consumables used in these processes are manufactured in developed countries and imported by developing countries via local suppliers. For this reason, import and transport fees as well as local supplier profit margins will increase the costs. Furthermore, developing countries may face difficulties in procuring NGS-related equipment and consumables due to supply chain limitations and import restrictions.

Once the genomic data is generated, data analysis and storage follow. The large volume of data generated by NGS requires substantial storage and computational capacity. Building and maintaining high-performance computing infrastructure can be challenging for developing countries due to the costs associated with this as well as the lack of expertize in maintaining such a computational system.

Lack of technical expertize

NGS requires highly specialized technical expertize, from sample preparation to data analysis. The shortage of trained personnel in developing countries can be a significant barrier.

For successful implementation of NGS, knowledge from a wide array of different fields is necessary. This includes skilled laboratory personnel for sample preparation and sequencing, bioinformaticians and statisticians for data analysis, and finally, clinical and molecular geneticists, molecular biologists, epidemiologists, etc. for data interpretation. While some of these fields may be well represented in developing countries, those skilled in bioinformatics are greatly lacking.

As a result of the lack of access to NGS infrastructure, there is a significant disparity between the number of skilled individuals to conduct NGS experiments in developing compared to developed countries. For developing countries to have the same level of participation in large-scale NGS projects that HICs have, more training initiatives are needed. This is challenging as many developing countries have limited training capacities in both education and human development. A few successful training initiatives have already been implemented in Africa. Training programs by genomics institutes, such as the KwaZulu Natal Research Innovation and Sequencing Platform (KRISP), the Center for Epidemic Response and Innovation in South Africa, and the Africa Center of Excellence for Genomics of Infectious Disease (ACE GID) in the Gambia, have had much success in training African scientists. Additionally, programs such as the Human Hereditary and Health in Africa (H3Africa) Bioinformatics Network (H3ABioNet) also provide training programs.[10]

Poor access to and representation on reference databases

As has been alluded to previously, there is a lack of genomic data representing populations in developing countries. This has led to the poor representation of these populations in reference databases and can make the interpretation of genomic data emanating from these regions very challenging. These reference databases are therefore biased toward populations from HICs. Furthermore, developing countries may lack the resources to build and maintain their own comprehensive reference databases.

Limited or overly restrictive regulatory and ethical challenges

NGS involves handling sensitive genetic information and navigating complex regulatory and ethical frameworks. Developing policies and frameworks governing the handling of genomic data has been seen in some developing countries, for example, South Africa, Zambia, Nigeria, Mexico, and India.[11-14] In contrast, most developing countries lack regulations/policies relating to genomic data. Initiatives such as H3Africa and the MX Biobank Project have been instrumental in fostering discussions regarding the importance of establishing such regulations going forward.

Funding and research prioritization

While recent initiatives have made a move toward funding research in developing countries specifically, there is still a lack of funding in these areas as compared to that in developed nations. This limited funding may lead to a lack of priority for NGS projects since the costs are so high. In turn, this will hinder progress in the field.

Funding that is given to developing countries has a tendency of being irregular, lack the ability to include overheads, and does not take into account the increased costs of performing research in developing countries. This and other relevant points are discussed in detail by van Helden.[15]

Addressing these barriers requires a multifaceted approach involving government support, international collaborations, capacity building, and efforts to reduce the cost of NGS technologies. Initiatives aimed at improving education, infrastructure, and regulatory frameworks can help LMICs overcome these challenges and harness the potential of NGS for healthcare and scientific research.

The way forward for NGS in LMICs—Potential solutions

The massive reduction in sequencing costs was unfortunately not accompanied by the reduction in the prices of most of the available sequencing instruments. Furthermore, the costs of reagents needed for NGS are also prohibitive. In most LMICs, reagents need to be imported which adds additional costs to the already high cost of the reagents. Added to this, many manufacturers of NGS kits and reagents use local distributors who add profit margin to the costs.[5]

While these challenges seem insurmountable, several measures could be implemented to overcome these hurdles. We highlight some potential solutions below.

Investment in local infrastructure

The lack of access to appropriate infrastructure is a major impediment to NGS access in LMIC. The scientific community in these countries needs to engage with their governments to increase investment in NGS infrastructure development. The need for these specialized laboratories was perfectly illustrated in the response to the COVID-19 pandemic. Genomics research was critical during the height of the COVID-19 pandemic to identify new SARS-COV-2 variants of concern. Interestingly, even though the number of reported COVID-19 cases was relatively similar in HICs and LMICs, the HICs submitted 10-fold more SARS-CoV-2 sequences per COVID-19 cases.[7] Therefore to be more prepared for future pandemics, investment in NGS in LMICs should be a priority.

In LMICs, academic and research institutions are ideally suited to establish centralized sequencing facilities to provide high-quality, cost-effective NGS services to several researchers. In South Africa, the South African Medical Research Council (SAMRC), in collaboration with the Beijing Genomics Institute and the South African Department of Science and Innovation designed and built a fit-for-purpose sequencing facility to provide NGS to researchers at an affordable cost (Fig. 5.2). The large government investment into building the infrastructure allowed for existing laboratories at the SAMRC to be renovated to suit the planned NGS workflows. The SAMRC Genomics Platform officially opened in July 2019 and has provided sequencing services ranging from whole genome sequencing (WGS) transcriptome sequencing (RNA-Seq), whole exome sequencing (WES), and whole genome bisulfite sequencing (WGBS) to researchers from South Africa and many other African countries.

Capital expenditure investment

Sustainable investment capital expenditure (capex) should also be a top priority when establishing a sequencing facility. Since 2005, the NGS market has grown significantly, with many companies providing sequencers with different costs, using varying chemistries, and producing varying amounts of data. Sequencers are large investments, and the choice of sequencing platform should be carefully considered based on the objective of the facility and the level of capex investment. Facilities specializing in sequencing the genomes of viruses and bacteria, for example, should consider smaller instruments producing less data suited for sequencing of small genomes, while facilities wanting to sequence large genomes (humans, animals, and pets) would need to invest in larger instruments.

The current NGS market is dominated by Illumina, which alone offers sequencing platforms for a wide range of sequencing applications suitable for any type of sequencing laboratory. ThermoFisher's Ion Torrent systems, while less accurate than the Illumina systems, are also still currently being used because of their affordability.[16] More recently, MGI entered the market, and its share of the market has

FIGURE 5.2 The South African Medical Research Council's genomics platform.

The laboratories contain state-of-the-art sequencers and automated extraction platforms.

From Craig Kinnear.

steadily increased over the past 3 years. Moreover, two companies, Pacific Biosciences (PacBio) and Oxford Nanopore Technologies (ONT), have pioneered ultra-long read sequencing. With all these options available the choice of platform suited to the objectives of the facility is not an easy decision. Things to consider when deciding on an instrument would be the cost of the instrument, the footprint of the instrument, the technical specifications, and the facility throughput requirements.

The cost of the instrument is one of the largest obstacles to overcome when establishing a sequencing facility. Depending on the instrument, one could purchase a sequencer for as little as $10,000 to as much as $1 million. Therefore the level of capex investment will be a major factor in acquiring a sequencer.

In addition to the sequencers, other peripheral equipment is also needed to prepare the sequencing libraries. These instruments should also be purchased and maintained so capex investments should also take these pieces of equipment into consideration.

Training programs

Investments in local NGS infrastructure are worthless unless one also invests in training personnel to work in these facilities. These training initiatives should focus on both the experimental and computational aspects of NGS and should ideally be supplemented by mentorship programs to enhance competency.[10] Additionally, the basic principles of genomics and bioinformatics should be integrated into the curricula of secondary schools with more advanced training being taught at undergraduate and postgraduate levels.[17,18] Such undergraduate training programs are already incorporated into life sciences courses in many developed countries that include approved degrees in bioinformatics.[19,20]

The availability of such accredited undergraduate degree programs makes it easier for students to transition from undergraduate to postgraduate level. In most African countries, these opportunities are not available, so aspiring researchers must rely on bioinformatics short courses for their bioinformatics training. While these courses are valuable and have been successful in providing training to many African researchers,[21] many lack structured critical mentorship programs.

Conclusion

NGS has revolutionized the fields of genetics and genomics and has driven precision medicine all over the globe. Moreover, we have seen how NGS can be harnessed in response to worldwide pandemics. It is therefore quite concerning that most NGS technologies still remain out of reach for many LMICs. The costs of sequencing, the inadequate infrastructure, poor access to and representation in public databases, and lack of training initiatives are among some of the challenges that LMICs face that hamper NGS implementation; however, these barriers to entry are not insurmountable. There is much genetic diversity on the African continent, and it remains in the interest of everyone invested in genomics that genomics in Africa (and in LMICs on other continents) flourishes. This can only be achieved if investments in infrastructure and training are made.

References

1. Lander ES, Linton LM, Birren B, et al. Initial sequencing and analysis of the human genome. *Nature*. 2001; 409(6822):860–921. https://doi.org/10.1038/35057062.
2. Carrasco-Ramiro F, Peiró-Pastor R, Aguado B. Human genomics projects and precision medicine. *Gene Ther*. 2017;24(9):551–561. https://doi.org/10.1038/gt.2017.77.
3. Williams SM, Sirugo G, Tishkoff SA. Embracing African genetic diversity. *Med*. 2021;2(1):19–20. https://doi.org/10.1016/j.medj.2020.12.019.
4. Wonkam A. Sequence three million genomes across Africa. *Nature*. 2021;590(7845):209–211. https://doi.org/10.1038/d41586-021-00313-7.
5. Helmy M, Awad M, Mosa KA. Limited resources of genome sequencing in developing countries: challenges and solutions. *Appl Transl Genomics*. 2016;9:15–19. https://doi.org/10.1016/j.atg.2016.03.003.
6. Reddy TBK, Thomas AD, Stamatis D, et al. The Genomes OnLine Database (GOLD) v.5: a metadata management system based on a four level (meta)genome project classification. *Nucleic Acids Res*. 2015; 43(1):D1099–D1106. https://doi.org/10.1093/nar/gku950.
7. Brito AF, Semenova E, Dudas G, et al. Global disparities in SARS-CoV-2 genomic surveillance. *Nat Commun*. 2022;13(1):7003. https://doi.org/10.1038/s41467-022-33713-y.
8. Jongeneel CV, Kotze MJ, Bhaw-Luximon A, et al. A view on genomic medicine activities in Africa: implications for policy. *Front Genet*. 2022;13. https://doi.org/10.3389/fgene.2022.769919.
9. Mulder N, Adebamowo CA, Adebamowo SN, et al. Genomic research data generation, analysis and sharing—challenges in the African setting. *Data Sci J*. 2017;16. https://doi.org/10.5334/dsj-2017-049.

10. Inzaule SC, Tessema SK, Kebede Y, Ogwell Ouma AE, Nkengasong JN. Genomic-informed pathogen surveillance in Africa: opportunities and challenges. *Lancet Infect Dis.* 2021;21(9):e281−e289. https://doi.org/10.1016/S1473-3099(20)30939-7.
11. Andanda P, Govender S. Regulation of biobanks in South Africa. *J Law Med Ethics.* 2015;43(4):787−800. https://doi.org/10.1111/jlme.12320.
12. Chanda-Kapata P, Kapata N, Moraes AN, Chongwe G, Munthali J. Genomic research in Zambia: confronting the ethics, policy and regulatory frontiers in the 21st Century. *Health Res Pol Syst.* 2015;13(1). https://doi.org/10.1186/s12961-015-0053-4.
13. Nnamuchi O. Biobank/genomic research in Nigeria: examining relevant privacy and confidentiality frameworks. *J Law Med Ethics.* 2015;43(4):776−786. https://doi.org/10.1111/jlme.12319.
14. Rojas-Martínez A. Confidentiality and data sharing: vulnerabilities of the Mexican genomics sovereignty act. *J Community Genet.* 2015;6(3):313−319. https://doi.org/10.1007/s12687-015-0233-5.
15. van Helden P. The cost of research in developing countries. *EMBO Rep.* 2012;13(5). https://doi.org/10.1038/embor.2012.43.
16. Berry M. Next generation sequencing and bioinformatics methodologies for infectious disease research and public health: approaches, applications, and considerations for development of laboratory capacity. *J Infect Dis.* 2020;221.
17. Karikari TK, Punta M. Bioinformatics in Africa: the rise of Ghana? *PLoS Comput Biol.* 2015;11(9). https://doi.org/10.1371/journal.pcbi.1004308.
18. Tastan Bishop O, Adebiyi EF, Alzohairy AM, et al. Bioinformatics education—perspectives and challenges out of Africa. *Briefings Bioinf.* 2015;16(2):355−364. https://doi.org/10.1093/bib/bbu022.
19. Nanjala R, Nyasimi F, Masiga D, Kibet CK, Ouellette F. A mentorship and incubation program using project-based learning to build a professional bioinformatics pipeline in Kenya. *PLoS Comput Biol.* 2023;19(3). https://doi.org/10.1371/journal.pcbi.1010904.
20. Sayres MAW. Bioinformatics core competencies for undergraduate life sciences education. *PLoS One.* 2018;13.
21. Mulder NJ, Adebiyi E, Alami R, et al. H3ABioNet, a sustainable pan-African bioinformatics network for human heredity and health in Africa. *Genome Res.* 2016;26(2):271−277. https://doi.org/10.1101/gr.196295.115.

Computational disease-risk prediction: Tools and statistical approaches

Emile R. Chimusa

Department of Applied Sciences, Faculty of Health and Life Sciences, Northumbria University, Newcastle, Tyne and Wear, United Kingdom

Introduction

Several forces including mutation, recombination, migration, geographic, and genetic isolation, population bottleneck, admixture, religion and culture, and natural selection have contributed to shaping the genetic diversity of human populations, yielding differences in diseases susceptibility, as well as variability in response to drugs and treatments. The primary goal of statistical genetics is to detect and characterize genetic variation to understand population diversity and evolution, as well as genetic risk factors that can be linked or associated with traits or human conditions.[1–3] Comprehensive knowledge of disease risk associations can inform clinicians or geneticists and assist in the prevention, diagnosis, prognosis, and treatment of diseases.[4]

Disease-scoring statistics such as genome-wide association studies (GWAS) have been successfully used as a common approach to detect variation in genes that are associated with a wide range of conditions.[1] However, the goals of GWAS have not been fully realized partly due to the disappointing and contradicting results that have been reported.[1,5] Particularly, in most of the traits, only a small fraction of the whole genetic contribution is attributable to variation at a specific locus of the genome.[5,6] For instance, height is known to be a highly heritable trait, yet the largest effect size that has been reported so far accounts for only approximately 1% of the genetic variance.[7] Similarly, in other traits, the missing heritability remains very large.[8,9] The fact that newer studies with larger sample sizes can include an increased range of allele frequencies means that novel variants that have never been associated with a phenotype are constantly identified. This suggests that human conditions are influenced by the additive or interactive effects of several loci with diminutive effects, which are very difficult to detect with current GWAS approaches.[7] These, in part, have made it difficult to understand and interpret the biological basis of the genetic variation for human conditions. That said, these results have pointed to the fact that most human conditions have a polygenic architecture, in which the variation and diversity are due to number of genetic variants across the genome, with each having a petite effect size.[7–9] This is supported by studies that aggregated the effects of several polymorphisms that are moderately associated with the conditions into a score, from which the score was found to be significantly associated with the conditions in different study samples.[10]

Population Genomics in the Developing World. https://doi.org/10.1016/B978-0-443-18546-5.00006-1

Today, thousands of genomes have been sequenced with millions of variants detected in the human genome. These genomic variations significantly contribute to a remarkable variability in human conditions' with major evolutionary and medical implications. However, most large-scale GWAS (some studies reaching up to millions of subjects) and population genetic structure studies have largely been conducted on individuals of European and/or Asian ancestry.[5] This disparity needs to be corrected to advance equitable and global health research in understanding evolution and to improve personalized and as well as predictive and preventative medicine.

In this chapter, we summarize the concept of genetic ancestry and risk prediction from a GWAS perspective. We then discuss potential approaches that can leverage other nongenetic information and omics profiles in the dissecting human variation, evolution, and estimate risk prediction from host GWAS. We conclude with a brief discussion of future research direction where further work is needed.

Computational tools for genetic ancestry and population structure

As illustrated in Fig. 6.1, different evolutionary forces including migration, admixture, religion, culture, population bottleneck, natural selection, and geographic and genetic isolation have played a critical role to present human genomic variation.[11,12] Genomic variation within and between human populations has been measured from the observed distributions of population allele frequencies.[11] Although it is complex to represent the interplay between these evolutionary forces, Fig. 6.1 displays a basic illustration of some of the forces that shape variation in human populations.

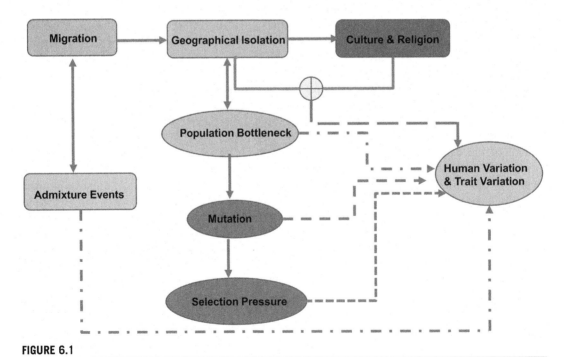

FIGURE 6.1

An example illustrating evolutionary forces toward human variation and diversity.

Over the past few decades, it has become clear that a mixture between diverged populations (admixture) has been a recurrent feature in human evolution.[11–13] During these admixture events, the individuals' chromosomes throughout subsequent generations contain a continuous breakdown of blocks inherited from ancestral populations. The admixed individual genome is viewed as a mosaic of segments of differing sizes of inherited genomic alleles from different ancestral populations. Rates of admixture continue to increase as long as migration and integration continue to occur in our society (Fig. 6.2). Admixed data provide a unique view into the genetic and environmental underpinnings of disease.[11,15]

Human diversity and genomic variation contribute to the reconstruction and understanding of population histories through computing or estimating genetic relatedness and/or divergence among individuals.[11,15] Elucidating genetic ancestry inference is crucial to better comprehend genomic variation patterns throughout human adaptive processes and evolution, as well as the implications in human health and heredity.[15] In fact, for decades, these patterns of human diversity has been identified and measured using approaches such as principal components analysis (PCA) and probabilistic approaches (admixture or ancestry inference also known as global or local ancestry inference [LAI]). PCA is a dimension reduction approach, from where genotypes or sequences data from each individual are projected along continuous axes of the variation.[15,16] Commonly, the first 2 or 3 axes of variation (eigenvector, also well-known as principal components), which are associated with the highest eigenvalues, are selected to define the most variability in the data set.

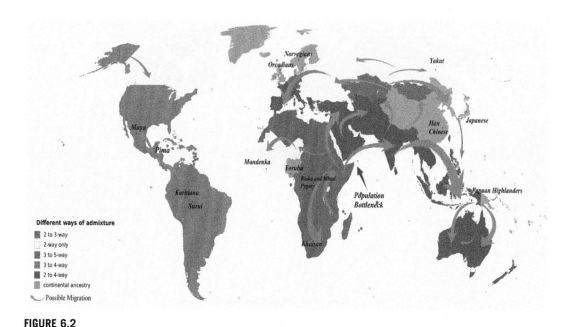

FIGURE 6.2

An indicative global map of increasing rate of mixed ancestry populations due to evolutionary forces.

Adapted figure from Chimusa ER, Defo J, Thami PK, Awany D, Mulisa DD, Allali I, Ghazal H, Moussa A, Mazandu GK. Dating admixture events is unsolved problem in multi-way admixed populations. Brief Bioinf. 2018;21(1):144–155.

Global ancestry (also known as genome-wide ancestry) is the overall genome average proportion of genetic contribution from an ancestral population, in contrast, local ancestry deconvolution yields more details on the population demographics and history through estimating the number of copies of an allele (0, 1, and 2) or chromosomal segment inherited from an individual's ancestry (ancestral population) at every position along the genome of an admixed individual. Global ancestry inference models or tools are divided into non-model-based or model-based.[13,15,16] Model-based methods make use of model parameters to compute allele frequencies and ancestral proportions, but non-model-based approaches estimate ancestry using algorithmic methods, they yield complex results and are harder to interpret, making their use less popular. LAI approaches are grouped into two,[15,16] either accounting for linkage disequilibrium (known as LD-based models) or the inability to model LD (known as non-LD-based models).[14,15] Advances in computational and inference approaches have enabled the development of over 20 ancestry inference models/tools used. These approaches have been improved over time to account for continental and closely related sub-continental ancestry, recent, complex, and ancient admixture events, or to infer local ancestry with neither statistical nor biological bias.[16] Today, no one tool can perform the best in all admixture scenarios. Fig. 6.3 provides an indicative illustration of commonly used global and local tools as they have been introduced over time.

Genomic variation contributes to the remarkable human trait variability with some variation having evolutionary and/or medical implications. However, in practice, there is a significant number of missing or untyped genomic markers through these generated human sequences, making it challenging to efficiently deconvolute human genomics variation in terms of ancestry inference and genomics risk assessment. Imputation of genotypes is a cost-effective plan to recover missing genomic markers in the

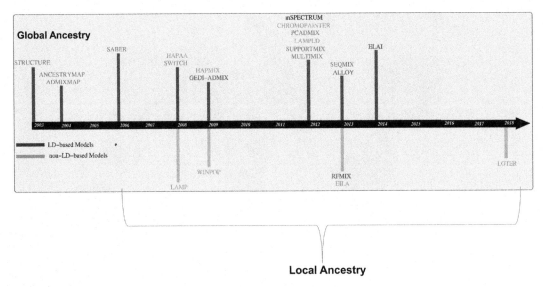

FIGURE 6.3

An illustration of most commonly used local and global ancestry inference tools.

From Geza E, Mugo J, Mulder NJ, Wonkam A, Chimusa ER, Mazandu GK. A comprehensive survey of models for dissecting local ancestry deconvolution in human genome. Brief Bioinf. 2019;20(5):1709–1724.

study sample using imputation tools.[17] The imputation of missing genotypes continues to play a significant role in increasing the power of genetic risk association studies by increasing the number of markers that can be investigated. Imputation approaches leverage known information including linkage disequilibrium (LD) between the missing genomic markers and typed flanking genomic markers. Approximately, 80 tools have been developed for genotype imputation (https://bioinformaticshome.com/tools/imputation/imputation.html#gsc.tab=0).[17] There is an unmet need to benchmark these tools in a wide range of diverse populations and reference panels to inform studies of the best tool that suits a given sampled population. However, it is apparent that imputation accuracy decreases with increasing genomic diversity, and this will have an impact on GWAS in multiway admixed and diverse populations. Genomic diversity, low LD, genetic mixture, and episodic selection pressure continue to challenge several population genomics methods such as global and LAI and imputation approaches; therefore, these challenges have subsequently constituted a challenge in identifying genetic variants with modest risk or protective effects during association studies.

Computational genome-wide association studies

The delineation of health and disease from association mapping holds promise to bridge the gap between clinical translation and statistical genomics with respect to disease risk prediction and association. This has hold great promise for improving diagnostics, screening, genetic testing, and counseling in global populations.[18–20] Variants associated with diseases first identified in populations of European ancestry do not necessarily replicate in diverse populations such as those commonly found in developing countries[21–23] due to several reasons such as confounding of sociodemographic factors and lifestyle across populations, linkage disequilibrium (LD), and differing genomic markers architecture. Genetic determinants of common and rare conditions and their effect sizes have also been shown to vary significantly between human populations,[19,21,22] particularly among African populations who are characterized by low levels of LD, high genetic diversity, recent selection pressures, and longer histories on a continent with environmental heterogeneity.[13,15,16,24]

Although efforts such as the development of the China Kadoorie Biobank,[23] Multi-Ethnic Global Array (MEGA),[24] Global Screening Array (GSA),[25] and the H3Africa array[26] have recently optimized genome-wide DNA microarrays in diverse populations, there are still more challenges in GWAS such as:

- Modest GWAS sample size in diverse populations.
- Stratification due to (a) the correlation of environmental exposures and ancestry, (b) multiwave admixture, (c) pre-/postadmixture selection pressure.
- Translation of associated genomic loci into suitable clinical and biological hypotheses.[21]
- Comprehension of how multiple moderately associated genomic loci within genes interact to dictate human condition variability.[21]

Although the above are challenges we face only on a daily basis as population geneticists and bioinformaticians, control of population stratification in GWAS is one of the biggest concerns, ensuring that observed associations are consistent with genetic effects of each genomic locus rather than just statistical correlations with ancestry.[27,28] For this reason, the mixed model becomes an adoptive and attractive approach in GWAS as it allows the inclusion of all samples irrespective of ancestry. Mixed model approaches contribute to addressing population stratification by modeling distant relatedness between sampled individuals due to ancestry correlation.[27,28] Several models have been implemented and we chronologically list some of them in Fig. 6.4A. Mixed models often produce greater statistical power, both through increased sample size and by controlling for the variance explained by the genetic relatedness between sampled individuals, that is, through a random effects part of the mixed model.[29,30] Several studies showed that variants of low-frequency (1%−5%) or at the boundary, may often attain genome-wide significance cut-off in mixed models due to the imperfect asymptotic and bias distribution that is often not well accounted for in most of the mixed models.[30] It is worth noting that mixed models may not fully control for subvariant structure between affected and unaffected samples, especially if there are nongenetics components to phenotypic associations with locus-specific ancestry due to admixture.[21,22,31] Nongenetic factors including sociodemographic, clinical parameters, and environmental exposures may be associated with genetic ancestry effects due to shared local, familial, or community backgrounds or due to the linkage between sociocultural factors and genetic ancestry.[32] There is an unmet need for effective model development to control for local ancestry specific tracts in GWAS, which may increase power and reduce false positives in mixed ancestry or multiancestry samples.[32,33]

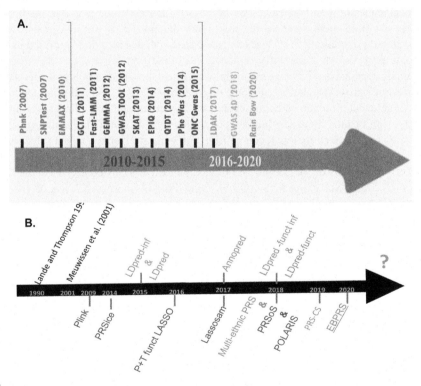

FIGURE 6.4

An example illustrating the chronological order of a partial list of (A) GWAS and (B) PRS tools.

No Permission Required.

The few studies that have leveraged local ancestry tracts in variant-level association analyses to African-Americans,[34,35] Latinos,[35] Hispanic cohorts,[36,37] and South Africa colored,[30] suggest added value beyond standard association scan.[38,39] Admixture association requires well-defined reference or correct proxy ancestral populations[40] to enable accurate LAI. Power can be optimized by combining association testing and admixture mapping;[41,42] however, this approach is rarely adopted because of the multistage steps required and the challenge posed in deconvoluting ancestry along the genome of complex multiway mixed ancestry individuals.[43] Therefore, there is a critical need to:

- Improve LAI accuracy.[15,16,34,42]
- Develop methods for optimizing the power of association testing[31,42,44] and polygenic risk scores (PRS) in admixed data.[19,42−44]
- Build integrative software for performing the multistage admixture analysis pipeline.[41]

In the next section, we will discuss the current approaches of PRS.

Computational artificial intelligence approaches for PRS
Origin of PRS

The concept of PRS was first introduced as a "molecular score" by Russell Lande and Robin Thompson,[45] entailing the sum of the additive effects of associated markers on human conditions. In 1990, the characterization of genomic markers was by restriction fragment length polymorphisms; at that time, the concept of GWAS followed by the selection of significantly associated genomic markers was yet to be introduced to the broad scientific community.[45] Russell Lande and Robin Thompson acknowledged that genome-wide LD between nearby genomic loci causally influencing variation between individuals could be exploited for selection. Later, in 2001, Meuwissen et al.,[46] introduced additional concepts and predicted the implications of the usage of dense genomic arrays. Methods to estimate genomic markers' effects sizes were considered, acknowledging issues of winner's curse, and issues of estimation, as the number of genomic markers is often larger than the number of sampled individuals.

Classification of current PRS methods

Methodologies for PRS construction differ mainly across two features: (a) how to decide which markers to include in the construction of PRS and (b) how weights utilized in the construction of PRS are generated. One of the critical issues at the early stage of the development of PRS models[47−55,57] was the fact that there was no inherent information regarding LD from summary statistics. Thus, if markers at a given genomic locus are in high LD with one another and are all included in the calculation of the score, then the score will potentially be inflated.[44] Therefore, a number of models have been introduced (Table 6.1) to control this, namely, LD pruning[48−50] and LD clumping.[49] These methods leverage Bayesian approaches to account for more information such as functional annotation through penalized or LASSO regressions. PRS models may be clustered based on how they model the LD and marker frequencies. Thus, we can group PRS models into two general classes: Bayesian PRS and non-Bayesian PRS approaches. Table 6.1 displays some popular and commonly used PRS tools, their models, and the year of publication (Fig. 6.5B).

Table 6.1 Summary descriptive of commonly used PRS tools.

Software	Model	Accounting for LD	Year	Comment	Effect size data format	Target format	Ref
PLINK	Standard PRS	LD clumping	2009	First tool to be applied in the calculation of PRS	Summary statistics	PLINK FORMAT	1[47]
PRSice	Standard PRS	LD clumping	2014	First standalone PRS software	Summary statistics	PLINK (bed/bim/fam)	1[48–50]
LDpred & LDpred inf	Bayesian PRS	Calculate LD from the reference panel	2015	First Bayesian PRS software	Summary statistics	PLINK (bed/bim/fam)	1[51]
AnnoPred	Bayesian PRS	Calculate LD from the reference panel	2017	Bayesian PRS	Summary statistics	PLINK (bed/bim/fam)	1[52]
Lassosum	Bayesian PRS	Calculate LD from the reference panel	2017	Penalized regression PRS	Summary statistics	PLINK (bed/bim/fam)	1[6]
PRSoS	Standard PRS	Spectral decomposition	2018	First PRS to implement spectral decomposition	Summary statistics	Oxford (.gen/.sample), variant call format (vcf)	1[53]
POLARIS	Standard PRS	Applies principal component analysis (PCA)	2018	First tool to apply PCA to adjust for LD	Summary statistics	PLINK (bed/bim/fam)	1[54]
LDpred-inf-funct LDpred-funct	Bayesian PRS	Calculate LD from the reference panel	2018	Bayesian PRS	Summary statistics	PLINK (bed/bim/fam)	1[55]
Multi-ethnic PRS	Bayesian PRS	Calculate LD from the reference panel and target data	2018	Bayesian regression PRS	Summary statistics	PLINK (bed/bim/fam)	1[56]
PRS-CS	Bayesian PRS	Calculate LD from the reference panel	2019	Bayesian regression PRS	Summary statistics	PLINK (bed/bim/fam)	1[57]
EBPRS	Bayesian PRS	No LD calculation and no training data	2020	Bayesian PRS	Summary statistics	PLINK (bed/bim/fam)	1[55]

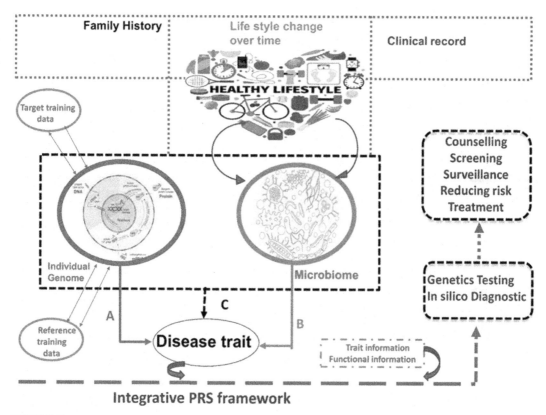

FIGURE 6.5

A suggested example for accounting for genetics and nongenetics factors in PRS.

Computing PRS

PRS approaches assume that alleles act additively, and consequently, a polygenic score can be constructed as the sum of risk alleles carried by a sampled individual.[4] PRS therefore summarizes the whole genome-wide data set into a single variable, which quantifies an individual's liability or tendency to a condition. As opposed to GWAS, which aims to identify an individual marker that is associated with a given condition, PRS aims at aggregating the genetic risk across the individual's genome into a score for each individual in the study. The score can then be utilized to evaluate the risk of the individual across conditions or traits. Although there are other methods[8–10,49] that have been introduced to summarize the risk of common markers that do not achieve the stringent GWAS cut-off, it is only PRS models that summarize these effects at the individual level.[46–55,57] Generally, PRS analysis is very simple to perform and very easy to interpret.[48–50,55]

The PRS calculation pipeline involves three steps: First, a GWAS is conducted either on an initial sample (more commonly known as the discovery sample) to obtain the summary statistics, or the summary statistics can be downloaded from one of the publicly available sources[46,49] such as the

GWAS catalog (www.ebi.ac.uk/gwas) or PRS can be obtained from the PRS catalog (https://www. pgscatalog.org). Trait-specific weights are then obtained from the summary statistics, including the nature logarithm of the odds ratio for binary conditions or the beta value for continuous traits. The second step for PRS analysis involves controlling the LD between the SNPs. Here, SNPs are either clumped, or pruned, or alternatively, some methods implement a Bayesian approach in modeling SNPs' weights from the summary statistics. Regardless of the method, the overall objective is to control the LD and avoid errors from an inflated score. The third step involves the construction of a PRS for each sampled individual in the study by multiplying the trait-specific weights (in the case of LD clumping and pruning approaches) or the posterior mean of the weights (in the case of Bayesian approaches) with the risk alleles an individual carries. While it is possible to utilize all markers in the construction of PRS, a *P*-value cut-off has been adopted by most tools in the selection of markers that show little or no evidence of associations. Consequently, PRS calculations are performed at various *P*-value cut-offs, and the most predictive cut-off is retained for downstream analysis.

As discussed above, common PRS models (Fig. 6.4B) begin by discarding markers in LD using an LD pruning approach and then applying a *P*-value cut-off in the selection of markers to be included in constructing PRS. This method, however, can discard several potential predictive markers during the pruning step, hence decreasing the predictive value and power. Nevertheless, this approach was the first used in some early studies that implemented PRS. There are however newer models that do not use summary statistics from GWAS, thresholds, or even take LD into considerations, as they include every marker in a GWAS (including imputed genotypes), and have been utilized to design polygenic risk models. To improve the predictive power and accuracy of PRS, a number of models have been introduced, including LD clumping as opposed to penalized regression methods, *P*-value-pruning, various Bayesian models that account for functional annotations, and approaches that leverage multiethnic polygenic risk scores by combining training data from European samples and the target population. Although these approaches have been widely investigated in recent years, almost all the available training data involve European samples; therefore, these methods have not been assessed using populations of non-European ancestry. This raises questions on their predictive power, transferability, and application across populations, and it is currently unclear how to reduce bias in predicting disease risk in populations of non-European ancestry.[10]

Opportunities and challenges in computational artificial intelligence disease-scoring statistics
Opportunities

PRS has several applications and may be utilized to deconvolute shared genomic etiology among human conditions and traits.[56,58] This is achieved through training the model score on one condition and testing the score against another condition. If there are associations, then there is a shared genomic basis. If shared genomic etiology has been detected, this suggests that a common molecular etiology exists between these conditions, which can be leveraged to detect individuals at risk and contribute to novel or optimal treatments. In addition, the shared genomic etiology may be leveraged to pinpoint problems with diagnosis. PRS is used to predict full disease risk with other nongenetics factors and thus can be utilized for stratifying individuals into different disease risk levels or groups. An example of this is in schizophrenia. PRS was developed and applied and showed that individuals whose PRS values fell within the upper 10th percentile had a 10-fold increased risk of developing schizophrenia than individuals whose PRS values fell into the lower 10th percentile.[8,9] Furthermore, in a recent study of breast cancer, individuals with the highest overall lifetime risk of 32.6% to develop breast cancer had PRS values that fell within the top percentile.[58–60] PRS or disease risk prediction has been applied as a guide to inform intervention or treatment options according to the risk clusters. However, the exceptional polygenicity of human traits makes unraveling mechanisms from PRS daunting.[13,24] To date, the development of PRS methods and tools,[47–55,57] their applications to understanding disease etiology, and their evaluation for clinical utility have been explored almost entirely in European ancestry and in developed countries[13], Fig. 6.4B illustrates partial PRS methods and recent tools.

Genetic diversity and its impact on PRS development and clinical interpretation—an opportunity for advancement

There is an unmet interest in understanding the dynamics of the origin of human variation, the evolution process, and its consequence in human conditions. However, current genomic risk assessment and disease risk prediction approaches have not been comprehensively benchmarked in large samples of admixed populations, particularly those from developing countries.[13,18] This raises the question as to how their clinical utility and real-world application can be made equitable across multiethnic populations[13,18] and, specifically, how to increase their predictability power in multiway admixed and diverse populations such. Harnessing the power of data sciences to design admixture polygenic risk score models tailored to diverse and multiway mixed ancestry populations is an unmet need to achieve powerful estimates and unbiased prediction power of disease risk prediction across diverse populations.[27,50,51,53]

Several recent studies showed that European ancestry-focused genetic studies provide little or diminishing returns to comprehend genetic etiology; however, a focus on multiethnic populations boosts our understanding of genetics and disease. Genomics data from multiethnic populations can enable a finer resolution of inference due to their high genetic diversity. Thus, mixed ancestry genomics are well-suited for tackling the challenge of understanding complex genetic etiology. Multiethnic populations are therefore a key present and future demographic yet they remain the most marginalized stratum of our current society. As the movement of people and their integration continue, the rates of admixture will continue to increase, highlighting the importance for their inclusion going forward, and ensuring accurate disease risk prediction inference is therefore crucial.

Observed current advances in human sequencing technologies (discussed in Chapter 5) and the development of computational tools can enable several next-generation PRS methods to account for the ancestry-sensitive and nongenetic factors to improve the predictive power of disease risk. These novel approaches can leverage various data from several current worldwide populations as ancestry proxies known to be involved in the migration and/or admixture processes yielding today's diverse and mixed ancestry populations around the globe (Fig. 6.5). Well-powered disease risk prediction in admixed and African-descent populations that leverage family history and environmental exposure will justify the implementation of public-based precision medicine and open avenues of investigation otherwise impossible. Delineating the disease risk according to individuals' PRS, family history of the disease, genetic ancestry, clinical covariates, and environmental exposure will solve current cross-populations PRS transferability issues. The variation in locus ancestry within mixed or multiethics individuals allows association between locus ancestry and phenotypes to be identified (*"ancestry-specific effects"*), and a comparison of locus effects of admixed and ancestral populations allows environmental interactions/correlations arising from the continent-of-origin to be identified (*"environment-modified effects"*).

PRS predictive power in real-world clinical populations—potential challenges and possible next steps

Besides its many applications, there are several challenges in the calculation of PRS, many of which have been discussed in this chapter.

We however believe that the delineation of PRS in association mapping will be the key in bridging the gap between statistical association and clinical translation in the following broad ways:

- Quantifying disease risk prediction attributable to combined genetics, host microbiome, and nongenetic factors (Fig. 6.2).
- Stratifying disease population risk, distinguishing ancestral from derived risk alleles. Considering the knowledge of the upper-limit risk stratification, population-specific disease risk methods predict the population-level risk of disease.
- Determining whether ancestry-specific PRS is needed for a given ethnic group or condition or pan-ethnic scores or for specific disease subtypes.
- Incorporating optimal selection criteria to select an accurate set of susceptibility loci due to the whole heritability can be distributed over several common/rare loci, each with petite effect sizes.
- Improving the precision of the model, as a function of sample size, in accounting for background/admixture LD, rare loci in the procedure for the estimation of coefficients, or weights, that will be attached to the selected loci.
- Detecting optimal delivery mechanisms, including cost-benefit and cost-effectiveness evaluations of action plans.
- Incorporating polygenic risk and possibly other predictive factors such as Electronic Health Records, to detect individuals at different risk levels for developing diseases (Fig. 6.5). This can be translated into increasing prevention by accounting for interventions according to the specific risk. Addressing organizational ethical and trustworthiness factors that affect risk perception, for their acceptance and adoption into real-world applications.
- Enabling the implementation of risk-based strategies of intervention; performing feasibility studies for action plans and, when possible, conducting randomized trials to directly benchmark the impact of new programs on health outcomes.
- Training health professionals in developing new risk communication tools.

To date, the performance and evaluation of genomic risk assessment scores have fallen short of the potential implied by heritability. When risk predictors exist based on sociodemographic or lifestyle factors, two key questions can be raised (1) to what extent can they be improved by including genetic information and (2) what is the potential of combined genetic and nongenetics risk scores (Fig. 6.5). These require the characterization of risk associated with individual factors and the exploration of potential interactions across these factors. For example, environmental factors such as lifestyle, diets, lifestyle, sociodemographic, and economic status are known as key drivers of human microbiome profile composition (Fig. 6.5). The sensitivity of the microbiome profiling not only as it relates to diet composition but also to food quantities, acting as dietary biosensors for the host, makes it unjust to neglect the role of the gut microbiome from consideration.[61]

We believe that including the host's microbiota profile in GWAS approaches for disease risk prediction may increase the predictive power and enhance clinical utility and real-world applications. This is because, despite empirical observations such as (a) the role of microbiome profiling in several human conditions, (b) the overlap between the association of microbiome profiling and genetic variation attributes to human conditions, and (c) the existence of heritable bacterial taxa, current GWAS approaches for heritability estimates do not account for the contributory role of the microbiome.

Conclusion

Predicting the individual risk of disease is currently an unrealized goal in predictive and preventative medicine as well as biomedical research as a whole. The delineation of PRS in association mapping is one of the keys to bridge the gap between statistical genomics and clinical translation. This chapter highlighted commonly used approaches and tools to dissect population genomics and genomic risk assessment. In addition, it provided a discussion on critical issues of equity in PRS research and the development of novel and integrative methods that may be tailored to all human populations, including multiethnic populations, particularly in developing countries who bear the greatest disease burden. Lastly, we highlighted the existing disparity between European and non-European populations in terms of PRS but highlight potential opportunities that will arise from sampling multiethnic populations.

Acknowledgments

We thank Peter O. Kimathi, Ephifania Geza and Jacquiline Mugo for figures and discussions. All computational analysis was performed through the Center of High-Performance Computing clusters (Cape Town, South Africa, https://www.chpc.ac.za/).

References

1. Bush WS, Moore JH. Genome-wide association studies. *PLoS Comput Biol.* 2012;8(12), 1002822.
2. Dudbridge F, Pashayan N, Yang J. Predictive accuracy of combined genetic and environmental risk scores. *Genet Epidemiol.* 2018;42(1):4−19.
3. Wray NR, et al. Polygenic methods and their application to psychiatric disorders and related traits. *J Child Psychol Psychiatry.* 2014;55:1068−1087.
4. Lewis CM, Vassos E. Prospects for using risk scores in polygenic medicine. *Genome Med.* 2017;9(1):96.
5. Yang J, Lee SH, Goddard ME, Visscher PM. GCTA: a tool for genome-wide complex trait analysis. *Am J Hum Genet.* 2011;88(1):76−82.
6. Mak TSH, et al. Polygenic scores via penalized regression on summary statistics. *Genet Epidemiol.* 2017; 41(6):469−480.
7. Robinson MR, Wray NR, Visscher PM. Explaining additional genetic variation in complex traits. *Trends Genet.* 2014;30(4):124−132.

8. Bulik-Sullivan BK, Loh PR, Finucane HK, et al, Schizophrenia Working Group of the Psychiatric Genomics Consortium. LD score regression distinguishes confounding from polygenicity in genome-wide association studies. *Nat Genet*. 2015;47(3):291.

9. Purcell SM, et al. Common polygenic variation contributes to risk of schizophrenia and bipolar disorder. *Nature*. 2009;460(7256):748–752.

10. Dudbridge F. Power and predictive accuracy of polygenic risk scores. *PLoS Genet*. 2013;9(3), 1003348.

11. Tishkoff S, Williams S. Genetic analysis of African populations: human evolution and complex disease. *Nat Rev Genet*. 2002;3(8):611–621.

12. Martin AR, Gignoux CR, Walters RK, et al. Human demographic history impacts genetic risk prediction across diverse populations. *Am J Hum Genet*. 2017;100(4):635–649.

13. Campbell MC, Tishkoff SA. African genetic diversity: implications for human demographic history, modern human origins, and complex disease mapping. *Annu Rev Genom Hum Genet*. 2008;9:403–433.

14. Chimusa ER, Defo J, Thami PK, et al. Dating admixture events is unsolved problem in multi-way admixed populations. *Briefings Bioinf*. 2020;21(1):144–155.

15. Geza E, Mugo J, Mulder NJ, Wonkam A, Chimusa ER, Mazandu GK. A comprehensive survey of models for dissecting local ancestry deconvolution in human genome. *Briefings Bioinf*. 2019;20(5):1709–1724.

16. Geza E, Mulder NJ, Chimusa ER, Mazandu GK. FRANC: a unified framework for multi-way local ancestry deconvolution with high density SNP data. *Briefings Bioinf*. 2020;21(5):1837–1845.

17. Das S, Forer L, Schönherr S, et al. Next-generation genotype imputation service and methods. *Nat Genet*. 2016;48(10):1284–1287.

18. Duncan L, Shen H, Gelaye B, et al. Analysis of polygenic risk score usage and performance in diverse human populations. *Nat Commun*. 2019;10(1):1–9.

19. Visscher PM, Wray NR, Zhang Q, et al. 10 years of GWAS discovery: biology, function, and translation. *Am J Hum Genet*. 2017;101(1):5–22.

20. Martin AR, Kanai M, Kamatani Y, Okada Y, Neale BM, Daly MJ. Current clinical use of polygenic scores will risk exacerbating health disparities. *Nat Genet*. 2019;51(4):584.

21. Chimusa ER, Dalvie S, Dandara C, Wonkam A, Mazandu GK. Post genome-wide association analysis: dissecting computational pathway/network-based approaches. *Briefings Bioinf*. 2018;20:690–700.

22. Shriner D, Adeyemo A, Ramos E, Chen G, Rotimi CN. Mapping of disease-associated variants in admixed populations. *Genome Biol*. 2011;12(5):223.

23. Chen Z, Chen J, Collins R, et al. China Kadoorie Biobank of 0.5 million people: survey methods, baseline characteristics and long-term follow-up. *Int J Epidemiol*. 2011;40(6):1652–1666.

24. Nelson SC, Romm JM, Doheny KF, Pugh EW, Laurie CC. Imputation-based genomic coverage assessments of current genotyping arrays: illumina HumanCore, OmniExpress, multi-ethnic global array and sub-arrays, global screening array, Omni2. 5M, Omni5M, and Affymetrix UK biobank. *bioRxiv*. 2017:150219.

25. Kalra S, Kaur RP, Ludhiadch A, et al. Association of CYP2C19*2 and ALDH1A1*1/*2 variants with disease outcome in breast cancer patients: results of a global screening array. *Eur J Clin Pharmacol*. 2018;74(10): 1291–1298.

26. Mulder N, Abimiku AL, Adebamowo SN, et al. H3Africa: current perspectives. *Pharmacogenomics Personalized Med*. 2018;11:59.

27. Korte A, Farlow A. The advantages and limitations of trait analysis with GWAS: a review. *Plant Methods*. 2013;9(1):1–9.

28. Korte A, Vilhjálmsson BJ, Segura V, Platt A, Long Q, Nordborg M. A mixed-model approach for genome-wide association studies of correlated traits in structured populations. *Nat Genet*. 2012;44(9):1066–1071.

29. Zhou X, Stephens M. Genome-wide efficient mixed-model analysis for association studies. *Nat Genet*. 2012; 44(7):821.

30. Chimusa ER, Zaitlen N, Daya M, et al. Genome-wide association study of ancestry-specific TB risk in the South African coloured population. *Hum Mol Genet.* 2014;23(3):796−809.

31. Seldin M, Pasaniuc B, Price A. New approaches to disease mapping in admixed populations. *Nat Rev Genet.* 2011;36:S21−S27.

32. McGrath JJ, Mortensen PB, Visscher PM, Wray NR. Where GWAS and epidemiology meet: opportunities for the simultaneous study of genetic and environmental risk factors in schizophrenia. *Schizophr Bull.* 2013; 39(5):955−959.

33. Awany D, Allali I, Dalvie S, et al. Host and microbiome genome-wide association studies: current state and challenges. *Front Genet.* 2018;9:637.

34. Marigorta UM, Rodríguez JA, Gibson G, Navarro A. Replicability and prediction: lessons and challenges from GWAS. *Trends Genet.* 2018;34(7):504−517.

35. Peprah E, Xu H, Tekola-Ayele F, Royal CD. Genome-wide association studies in Africans and African Americans: expanding the framework of the genomics of human traits and disease. *Public Health Genomics.* 2015;18(1):40−51.

36. Gonzalez S, Gupta J, Villa E, et al. Replication of genome-wide association study (GWAS) susceptibility loci in a Latino bipolar disorder cohort. *Bipolar Disord.* 2016;18(6):520−527.

37. Chen W, Brehm JM, Manichaikul A, et al. A genome-wide association study of chronic obstructive pulmonary disease in Hispanics. *Ann Am Thoracic Soc.* 2015;12(3):340−348.

38. Geza E, Mulder NJ, Chimusa ER, Mazandu GK. FRANC: a unified framework for multi-way local ancestry deconvolution with high density SNP data. *Briefings in Bioinformatics.* 2020;21(5):1837−1845. https://doi.org/10.1093/bib/bbz117, 5612160.

39. Chimusa ER, Mbiyavanga M, Mazandu GK, Mulder NJ. ancGWAS: a post genome-wide association study method for interaction, pathway and ancestry analysis in homogeneous and admixed populations. *Bioinformatics.* 2016;32(4):549−556.

40. Chakraborty R, Weiss KM. Admixture as a tool for finding linked genes and detecting that difference from allelic association between loci. *Proc Natl Acad Sci USA.* 1988;85(23):9119−9123.

41. Shriner D, Adeyemo A, Rotimi CN. Joint ancestry and association testing in admixed individuals. *PLoS Comput Biol.* 2011;7(12). e1002325.

42. Thornton TA, Bermejo JL. Local and global ancestry inference and applications to genetic association analysis for admixed populations. *Genet Epidemiol.* 2014;38(S1):S5−S12.

43. Marnetto D, Pärna K, Läll K, et al. Ancestry deconvolution and partial polygenic score can improve susceptibility predictions in recently admixed individuals. *Nat Commun.* 2020;11(1):1−9.

44. Coram MA, Fang H, Candille SI, Assimes TL, Tang H. Leveraging multi-ethnic evidence for risk assessment of quantitative traits in minority populations. *Am J Hum Genet.* 2017;101:218−226.

45. Lande R, Thompson R. Efficiency of marker-assisted selection in the improvement of quantitative traits. *Genetics.* 1990;124(3):743−756.

46. Meuwissen THE, Hayes BJ, Goddard MR. Prediction of total genetic value using genome-wide dense marker maps. *Genetics.* 2001;157(4):1819−1829.

47. Purcell S, Neale B, Todd-Brown K, et al. PLINK: a tool set for whole-genome association and population-based linkage analyses. *Am J Hum Genet.* 2007;81(3):559−575.

48. Choi SW, Mak TSH, O'reilly P. A guide to performing polygenic risk score analyses. *bioRxiv.* 2018:416545.

49. Euesden J, Lewis CM, O'Reilly PF. PRSice: polygenic risk score software. *Bioinformatics.* 2015;31(9): 1466−1468.

50. Choi SW, O'Reilly PF. PRSice-2: polygenic risk score software for biobank-scale data. *GigaScience.* 2019; 8(7):giz082.

51. Vilhjálmsson BJ, Yang J, Finucane HK, et al. Modeling linkage disequilibrium increases accuracy of polygenic risk scores. *Am J Hum Genet.* 2015;97(4):576−592.
52. Shi J, Park JH, Duan J, et al. Winner's curse correction and variable thresholding improve performance of polygenic risk modeling based on genome-wide association study summary-level data. *PLoS Genet.* 2016; 12(12). e1006493.
53. Chen LM, Yao N, Garg E, et al. PRS-on-Spark (PRSoS): a novel, efficient and flexible approach for generating polygenic risk scores. *BMC Bioinf.* 2018;19(1):1−9.
54. Baker E, Schmidt KM, Sims R, et al. POLARIS: polygenic LD-adjusted risk score approach for set-based analysis of GWAS data. *Genet Epidemiol.* 2018;42(4):366−377.
55. Marquez-Luna C, Gazal S, Loh PR, Furlotte N, Auton A, Price AL, 23 and Me Research Team. Modeling functional enrichment improves polygenic prediction accuracy in UK Biobank and 23andMe data sets. *bioRxiv.* 2018:375337.
56. Márquez-Luna C, Loh PR, , South Asian Type 2 Diabetes (SAT2D) Consortium, SIGMA Type 2 Diabetes Consortium, Price AL. Multiethnic polygenic risk scores improve risk prediction in diverse populations. *Genet Epidemiol.* 2017;41(8):811−823.
57. Hu Y, Lu Q, Powles R, et al. Leveraging functional annotations in genetic risk prediction for human complex diseases. *PLoS Comput Biol.* 2017;13(6). e1005589.
58. Ge T, Chen CY, Ni Y, Feng YCA, Smoller JW. Polygenic prediction via Bayesian regression and continuous shrinkage priors. *Nat Commun.* 2019;10(1):1−10.
59. Song S, Jiang W, Hou L, Zhao H. Leveraging effect size distributions to improve polygenic risk scores derived from summary statistics of genome-wide association studies. *PLoS Comput Biol.* 2020;16(2). e1007565.
60. Mavaddat N, Michailidou K, Dennis J, et al. Polygenic risk scores for prediction of breast cancer and breast cancer subtypes. *Am J Hum Genet.* 2019;104(1):21−34.
61. Mazidi M, Rezaie P, Kengne AP, Mobarhan MG, Ferns GA. Gut microbiome and metabolic syndrome. *Diabetes Metabol Syndr Clin Res Rev.* 2016;10(2):S150−S157.

Genotype versus phenotype versus environment

Elouise Elizabeth Kroon[1], Yolandi Swart[1], Roland van Rensburg[2], Katelyn Cuttler[1], Amica Corda Müller-Nedebock[1,3] and Maritha J. Kotze[4]

[1]*South African Medical Research Council Centre for Tuberculosis Research, Division of Molecular Biology and Human Genetics, Faculty of Medicine and Health Sciences, Stellenbosch University, Cape Town, South Africa;* [2]*Division of Clinical Pharmacology, Department of Medicine, Faculty of Medicine & Health Sciences, Cape Town, South Africa;* [3]*Stellenbosch University, South African Medical Research Council/Stellenbosch University Genomics of Brain Disorders Research Unit, Cape Town, South Africa;* [4]*Faculty of Medicine and Health Sciences, Stellenbosch University and National Health Laboratory Service, Tygerberg Hospital, Division of Chemical Pathology, Department of Pathology, Cape Town, South Africa*

Introduction

There are around 8 billion people in the world, each unique with visible and some not so visible differences. These differences or distinguishing characteristics are known as traits. The development of DNA sequencing technologies (discussed in Chapter 5) and statistical models (discussed in Chapter 6) has helped us to understand the genetic underpinnings of a combination of one or more of these traits, also known as a phenotype. Humans share approximately 99.9% of their DNA with any two genomes differing by approximately 0.1% of the nucleotide sites, accounting in part for the unique phenotypic variation observed between persons.[1] This 0.1% is not negligible and equates to roughly one variant per 1000 bases on average which can differ between individuals.[1] However, the genotype alone cannot sufficiently explain all observed phenotypic variations. Since the advent of genetics, the key question has been to determine the contribution of the underlying factors influencing phenotypic expression, which can vary from explicit distinctions such as eye color to more finely nuanced immune differences in disease susceptibility or resistance.

This phenotypic variance can be determined by heritable components (genotypic variance) and nonheritable components (environment). For a long time, a false dichotomy drove the idea that either nature or nurture was responsible for phenotypic expression. Today we know that both the genotype and environment are key for the understanding of phenotypic variation, and they should not be interpreted in isolation.

Genotype versus environment

The underlying traits (inherited characteristics) of a phenotype can be determined by a single gene (monogenic), which is usually rare, or due to multiple genes (polygenic). Mendelian genetics was

Population Genomics in the Developing World. https://doi.org/10.1016/B978-0-443-18546-5.00007-3

sufficient to model monogenic inheritance of discrete traits; however in the early 1900s, it became clear that this could not be applied to continuous traits.[2,3] As the distribution of phenotypic variation increases, the number of loci associated with a trait becomes more complex. Quantitative genetics aims to quantify the complexity of these effects of multiple interacting loci and environmental interactions with alleles on phenotypic variation.[3] The field developed from initially only partitioning the total phenotypic variation into the contributing genotypic and environmental components, to developing techniques to identify the phenotype-associated loci using quantitative trait loci (QTL) and genome-wide association studies (GWAS).

An example of quantitative methods can be seen in twin studies. Genetic studies comparing observable traits between monozygotic twins (MZ) and dizygotic twins (DZ) have added immense value to understanding the missing link between genotype and phenotypic expression, i.e., the environment. Newman recognized in some of the earliest twin studies that even seemingly "identical" or MZ who share a common genotype and who were raised in the same environment are never completely alike phenotypically.[4] In comparison DZ share on average 50% of their genes. Early classical twin studies aimed to describe the proportion of variation of a trait in the studied population due to inherited genes (heritability) compared to the environmental contribution. Traditionally the twin model captured the additive effect of genes (A) and shared environment (C) as well as nonshared environmental aspects specific to a person (E) and is commonly referred to as the ACE model. Classical twin models assume a perfect correlation for common environmental exposure on a specific trait, also known as the equal environment assumption. This assumption has been widely contested and is arguably a possible confounder for the larger heritability estimates found in twin studies.[5] Genetic twin studies demonstrated that from conception onwards, environmental factors as well as genotype-environment interactions (GEI) play a role in the expression of phenotypes.[6–8]

As an illustrative case: A young male patient in his thirties was referred for a thrombophilia genetic screen based on the diagnosis of a pulmonary embolism (PE). He had a known smoking history while his twin brother did not smoke and remained healthy Fig. 7.1. The smoking was identified as an environmental trigger for the genetic predisposition of inherited thrombophilia. This finding provides an example of personalized medicine where thousands of persons do not need to be tested to demonstrate the GEI within this clinical context when using the interpretation algorithm described by Kotze.[9]

Twin studies performed after emergence of the coronavirus 2019 disease (COVID-19) showed that multiple pathogen exposure together with other environmental factors (EF) may result in a unique individual immune response.[10] Postacute sequelae of COVID-19, generally known as long COVID, was only observed in one of the MZ cases referred for whole exome sequencing (WES), despite sharing the same genetic makeup.[10] This finding highlighted the significance of both host genetics and immune response in the severity and progression of the disease course.[10] Neither of the brothers had any known health conditions or comorbidities detected at a high prevalence in COVID-19 patients with severe disease, including Long COVID in South Africa and elsewhere.[11]

Today we understand that most traits are more complex and affected by gene, environment, gene-gene and GEI Fig. 7.2. The development of improved and cost-reduced sequencing platforms has led to the increased availability of genomic data. Despite this, many disease-causing variants are still unknown. Meta-analysis of data using GWAS has added to the statistical strength to identify susceptibility loci for multiple complex diseases. For noncommunicable diseases (NCDs) such as

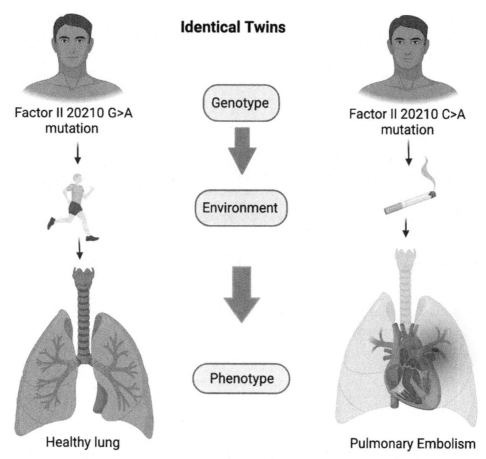

FIGURE 7.1 Differential response to smoking in identical twins with the factor II 20210 G > A single nucleotide variant.

Both identical twins have the genetic predisposition for inherited thrombophilia. However, this was only triggered by environmental exposure to smoking in the one twin.

Created in Biorender.com

cardiovascular disease (CVD), most of the identified loci are unable to fully account for the total heritability of disease outcome, possibly due to unidentified population specific variants, rare and high penetrance alleles, or GEI.[12,13] GEI are difficult to study since different statistical disease models and small sample sizes are often used making reproducibility increasingly difficult. In addition, most complex diseases are caused by gene variants with small effect sizes.[14]

Added to the complexity of identifying disease-causing variants, researchers still face multiple challenges to characterize the functional impact of genetic variation.[15] Although genetic variation contributes to many differences, not all genes are always expressed and so variants alone are not

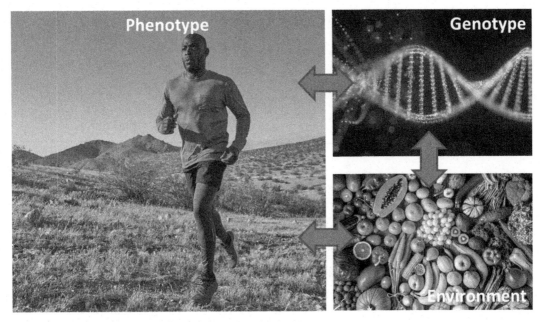

FIGURE 7.2 The genotype-phenotype-environmental triad.

The complexity of phenotypic expression is highlighted by genotype, environmental and genotype-environment interactions.

Images from Word stock images that does not require permission.

responsible for what can be observed. Epigenetic factors may disrupt the expression of genes in coding regions of exon sequences prior to transcription. In addition, miRNAs may regulate gene expression through posttranscriptional modification. Variants in noncoding regions are even more challenging to interpret and impact the transcription of coding genes.[15] Finally, the translation of proteins and protein assembly represent definitive processes that may disrupt the genotype-phenotype relationship.

Likely mechanisms in which the environment can impact this genotype-phenotype discordance, include epigenetics, miRNA alterations, protein misfolding, mitochondrial effects or transporter activity.[8] Combined environmental effects are as complex as the traits they affect and are not limited to factors external to the human body e.g., food, light, temperature, pollution, toxins, drugs, or even other social or economic factors. Internal physiological factors such as hormones, inflammation, metabolism, etc. also play a fundamental role. These EF may influence the expression, timing, presentation and severity of a trait or disease. The relative effect of EF on disease may occur at the population, individual or a molecular level.[16] It is hard to measure or record all variable EF which may impact gene expression. The importance of each factor would differ depending on the phenotype of interest and are at times dependent on an individual person's recollection of detail, especially with regards to factors of a more sensitive nature. For example, persons are not always forthcoming on the exact amount of alcohol they use. Not all relevant variables are therefore always recorded during participant recruitment or sampling.

Phenotype

An important research question emerging from the above-mentioned information is determining which traits are needed to quantify genotypic data and GEI. Multiple descriptors such as race and ethnicity are becoming increasingly redundant with the availability of recent advances in next-generation sequencing (NGS) technologies, which require a paradigm shift in thinking, as discussed below.

The trap of the observable

Key aspects of phenotyping related to individual genome and/or population genetic background may result in the trap of the observable. To date, genetic research has been inundated with redundant phenotypic descriptors. An example of such a redundant phenotypic descriptor is the term "race". It is often used interchangeably with ethnicity, despite having distinct meanings.[17,18] Race is a social construct and is used to characterize individuals into groups based on outward appearances (such as skin color, hair, eye color, etc.).[19] Ethnicity refers to the cultural and traditional heritage, as well as the language and social practices of particular groups.[19] Neither has a biological basis and is often used for socio-political agendas. Despite numerous calls and reviews to avoid the use of these classifiers in medical and scientific research, it is still being used regularly in research settings as covariates to account for environmental factors.

There is a long history of stratifying by self-identified race/ethnicity in genomic studies. The scientific and medical communities have used race to describe disparities in disease prevalence and outcome. Population groups were historically separated by geographical barriers (deserts, mountains, oceans, etc.), which resulted in the differential distribution of allele frequencies between population groups due to inbreeding within the group only.[20] Self-identified race became necessary to stratify a group of individuals based on a person's outward appearance and geographical location. Given the limited knowledge of and access to genotypic technology a decade ago, self-identified race was useful to help determine the appropriate treatment regimens for patients. For example, in genetically homogeneous groups such as the Ashkenazi Jewish population, a limited number of causative cancer variants may predominate, alerting clinicians to timely cancer screening and counseling. Furthermore, the spectrum of pathogenic variants in the two major breast and ovarian cancer susceptibility genes, BRCA1 and BRCA2 differ between populations based on their genetic structure.

Despite most of the science community refraining from using the term race, ethnicity is still used by some to describe participants' self-identified population groups. The predicament is that persons could harbor a disease-causing variant on a gene from an ancestry they did not self-identify as. The genetic variants/regions associated with certain phenotypic traits may be more prevalent in specific ancestry groups, as evidenced for the most common missense variant (HFE C282Y) responsible for the high prevalence of autosomal recessive hereditary hemochromatosis in the European population.[21] Although this genetic variant may be more prevalent in Europeans is does not preclude, variant detection and disease development in other population groups. Using self-identified race can therefore be problematic for disease diagnosis, biomarker development, and therapeutic targets.

All populations represent a long history of migration and mixing.[20] Reproductive isolation coupled with admixture (due to migration) gave rise to admixed individuals with distinct allele frequencies. Today, various populations around the world are still experiencing admixture events that not only affect their current but also future genetic composition. Based on this multiple racial categories will need to be continually created and updated as new allelic combinations are created by admixture events. It will become increasingly difficult to assign a race based on outward appearance or self-identified race/ethnicity. The observation of physical attributes is limited by the perception of the researcher, while self-identified race is limited by the person's self-perceptions and may be influenced by society's labels and categories. Ultimately, there is much ambiguity and genetic variation between current race classifiers. To rely purely on outward appearance to capture the differences would neglect to benefit all populations worldwide.

When self-identified race is used as a proxy covariable to account for socio-economic status in genomic association studies, there is a danger of potential harm caused by linking race as a sole determinant of standard of living. Social factors are key contributors to disease prevalence, but there is evidence that these factors do not account for all health differences. Some have argued that negative environmental, social and economic conditions during early years of life in African Americans detrimentally affect biological conditions in later life.[22] Although this assumption highlights many of the health disparities experienced by African Americans, it is important to consider that there is also extensive genetic individual variability in these environments within the African American population. Certain people will identify as African but harbor a substantial quantity of European ancestry or two people of European descent may share more Asian than European ancestry.

It is not uncommon to find multiple admixed individuals worldwide Fig. 7.3. For example, most individuals residing in southern Africa (South Africa, Namibia, Botswana, Lesotho, and Eswatini) are mainly unaware of the admixed nature of their genome.[23,24] In Southern Africa, some individuals can be up to five-way admixed (having contributions from five ancestral populations) with ancestral contributions from multiple populations with the same socio-economic status. Examples include the spectrum of pathogenic variants detected in the low-density lipoprotein (LDL)-receptor gene underlying Familial Hypercholesterolemia in the South African population, due to a founder effect or recurrent mutational events.[21,25] Multiple admixture is not only found in southern African citizens, but also in other populations worldwide who received ancestral contributions from unknown ancestors (e.g., Peruvians from Brazil).[26] Researchers should therefore be cautious to use a single population descriptor such as African Americans to represent all African populations in genomic association studies.

Genetic admixture is the result of recent interbreeding between previously separated populations. The result is a genome which contains a mosaic of previously isolated populations with heterogeneity by ancestry in allelic effect size.[27] Admixed individuals genetic ancestry can differ substantially, with their global ancestry proportions ranging from nearly 100% to as low as 20% although they self-identify as the same ethnic group. At first glance this might seem insignificant, however the local genetic ancestry can vary drastically, even between individuals that have relatively similar global ancestry proportions of the same ancestry. The more fine-scaled ancestry is known as local ancestry (LA).[27,28] Estimating LA is specifically important for gene regulation, which occur in discrete nearby locations to the gene, such as transcription star sites or tagging single nucleotide polymorphisms (SNPs) of micro array chips.

The inclusion of ancestral proportions (and by implication ethnicity) in statistical analysis is required to adequately correct for population substructure in admixed individuals. The inference of genetic ancestry is founded on the knowledge of the dispersal of diversity among human populations that mirrors the evolutionary and demographic history of our species. Genetic ancestry estimation can highlight the nuances of ancestry and dispel the notion of "race" in humans as well as the practice of

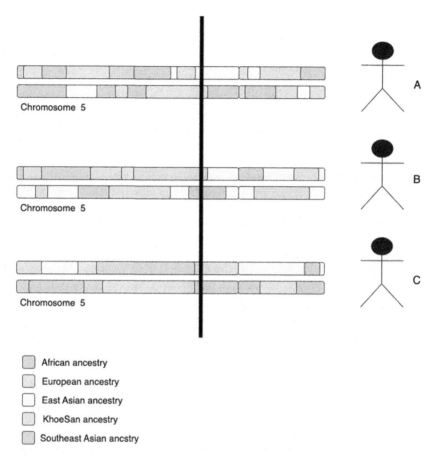

FIGURE 7.3 Complex nature of multiway complex admixed genomes.

For illustrative purposes the disease-causing variant of a specific disease can be located on the same genetic region (indicated with the black line) on chromosome 5 for all three multiway admixed individuals (A, B, and C). This could be missed in genomic association studies. Relying only on the outward appearance or self-identified "race" risks misclassifying the ancestry present at the disease-causing variant. All three persons may self-identify as being of the same "race" but may have inherited a different local ancestry at the specific region.

equating "race" with ancestry. It is equally capable of giving credence to the unscientific idea that humans subdivide into distinct biological races and implying that there are clear-cut connections between DNA and specific geographic regions or ethnic groups.

Genetic ancestry inference is ultimately a statistical procedure that depends on analytical software and genetic markers obtained from genotypic data (albeit whole genome sequencing data or gene chips array data).[29] Without an individual's accurate ancestral proportions, inaccurate associations will be made in terms of the genetics of disease susceptibility, with resulting false associations leading to wasting valuable laboratory resources. In addition, early screening and personalized treatment for most populations worldwide will not be feasible.

Before tailored genetic testing based on the unique ethnic landscape of the community can be implemented, the inference of genetic ancestry in complex multi-way admixed individuals needs to be accurate. In turn, this enables the development of cost-effective first-tier genetic tests before extending the analysis to NGS analysis in genetically uncharacterized cases. In South Africa research performed over more than 3 decades evolved into the development of a BRCA1/2 point-of-care (POC) assay applicable across ethnic groups, and irrespective of family history or tumor subtype.[24]

To best illustrate the unintended consequences of the inappropriate use of racial categorization in genetic studies, we offer further examples from diabetes and vitamin D supplementation:

Glycated hemoglobin (HbA1c) is a screening tool for glycemic control in diabetic patients. HbA1c is the reference standard used to diagnose and measure the long-term (red blood cells with a lifespan of 120 days) glycemic control in diabetes patients.[30] The long-term management of diabetes is important to prevent the possible adverse effects of hyperglycemia. Multiple factors influence the relationship between blood glucose levels and HbA1c, including differences between race/ethnicity and several hemoglobin variants.[31]

Variable results of HbA1c were observed in individuals of mixed ancestry in South Africa. HbA1c generally performed poorer in nonobese individuals and therefore HbA1c may be a suboptimal test for uncontrolled glucose screening in a large proportion of the African population.[32] So far, there have been limited investigations of genetics as a potential factor explaining these HbA1c related racial differences. It was first thought that the differences seen in HbA1c between races were due to health disparities; however, new research suggests that there may be ancestry-specific genetic variants affecting the erythrocyte biology in individuals of African descent which is partly due to selective pressure to resist malaria.[30] This observation was also evident in a study investigating African Americans and indicated that despite having comparable glucose levels, "blacks" had significantly higher HbA1c levels than non-Hispanic "whites".[33] Furthermore, "black" individuals with the risk associated A allele of rs334 in the HBB gene, also known for sickle cell trait (SCT) and hemoglobin S, had higher HbA1c levels.[33] This was confirmed in a study conducted by Skinner et al. in a Senegalese population.[34]

The commonly used threshold for diabetes of ≥6.5% HbA1c may, therefore, substantially misclassify the status of many African people with poor glycemic control. The impact of the screening and monitoring of diabetic patients with African ancestry is enormous. It is recommended for individuals of African descent that HbA1c should be used in combination with another diagnostic tool such as fasting blood glucose levels, fructosamine, or glycated albumin to decrease the chances of misdiagnosis in individuals harboring African ancestry. However, mixed "race" makes this inherently difficult since a person may appear "nonblack" based on their outward appearance, but potentially harbor Africans-specific genetic variants in certain regions of their genome.[30] Consequently, such a case may be misdiagnosed as diabetic or nondiabetic and will not receive the correct recommended treatment. In this context, it is extremely important to understand the difference between reference intervals and clinical decision limits considered to support the interpretation of numerical clinical pathology test results.[35] The typical distribution of results from a healthy reference population should not be confused with the cut-off levels suitable to inform the need for patient follow-up aimed at timely intervention.

Therapeutic approaches that are designed based on "racial categorization" (simple black/white) or self-identified ethnicity, may lead to adverse drug reactions. Nutritional needs and the benefit or harm from food supplementation also vary depending on genetic factors as well as age and health status. The recommended clinical decision limit for vitamin D, for example, may vary depending on the specific laboratory method used.[36] It was also found that persons of African ancestry typically have lower than the recommended optimal levels of serum 25-hydroxyvitamin D_3 (25(OH)D) due to increased levels of melanin pigmentation in their skin.[37,38] Vitamin D is not a natural occurring vitamin in the body and is made from 7-dehydrocholesterol via the effect of sunlight on the skin. Cholecalciferol or better known as vitamin D-3 (the intermediate), is then converted to 25(OH)D by the liver.[38] The cutaneous synthesis of vitamin D depends on the UV-B spectrum in sunlight. This process is efficient in darker skin and in northern latitudes with less UV-B. Vitamin D deficiency is characterized as levels below a recommended threshold of 12 ng/mL (<30 nmol/L) excluding consideration of race/ethnicity or skin color. This may lead to persons of an African descent to be predominantly classified as vitamin D deficient. Setting the recommended threshold of vitamin D deficiency to a certain threshold based on European ancestry, risks considering other ancestries (not limited to Africans) as either deficient or give vitamin D supplementation that may be harmful.[38] However, consensus has not been reached about the serum concentrations of 25(OH)D required for overall health. To maintain bone health and normal calcium status, the focus should rather be to aim for a serum 25-hydroxyvitamin D [25(OH)D] concentration of at least 30 ng/mL (75 nmol/L) as recommended in the Endocrine Society Clinical Practice Guidelines, but below 50 ng/mL (125 nmol/L) to prevent adverse effects.[39]

Lower levels of vitamin D are associated with decreased levels of bone mineral density (BMD) and have an increased risk of bone fractures and osteoporosis, mostly in Europeans. Vitamin D supplementation may aid in preventing or reducing rates of bone loss, since higher serum 25(OH)D levels are associated with higher BMD. However, studies reported no reduction in bone loss in postmenopausal African American women with vitamin D_3 supplementation compared to placebo.[37] Interestingly, maintaining the recommended serum 25(OH)D levels (>30 ng/mL) by the Endocrine Society guidelines, had no additional benefit in prevention of involutional decline in bone strength. The question whether the same lower observed serum 25(OH)D levels are recommended for individuals of mixed ancestry is mostly unknown. There may be a different daily recommended threshold of 25(OH)D required for multiple complex admixed individuals from South Africa.

The complexity of the association between vitamin D and osteoporosis cannot only be explained by self-identified race and/or genetically inferred ancestry. Multiple factors need to be simultaneously interpreted and pathology-driven decision-making can act as an intermediate phenotype to close the current gap in genomic association studies. Okunola et al. (2023) investigated the implication of aligning vitamin D levels with genomic data for differential diagnosis of familial versus lifestyle and/or therapy-induced disease manifestation.[40] This study used an "extremes of outcome" PSGT strategy which allows for the reuse of whole exome/genome sequencing data, thereby reducing research costs associated with large population studies performed outside the reality of clinical practice. The value of data sharing was evidenced by prior detection of aromatase inhibitor-induced bone loss linked to the CYP19A1 gene in the same study population,[41] which was contrasted in this study with a significant association between osteoporosis at baseline and functional polymorphisms in the vitamin D receptor (VDR) gene. Only using self-identified race as a phenotypic classifier may result in suboptimal treatment for bone loss in postmenopausal women of African descent.

Disease risk differs between populations worldwide not only because of genetic differences, but also due to complex gene-environment interactions. However, defining populations or races in genetic studies has been neglected. This led to much confusion and ongoing misuse of race/ethnicity in genomic association studies. Nonetheless, previous studies on self-identified race/ethnicity laid the foundation and prompted scientists to develop databases and software to be able to infer ancestry at each genomic region with higher accuracy. To investigate certain patterns that influence disease pathology and treatment, it will be necessary to consider the ancestry at each genomic region of an individual regardless of outward appearance. The correct use of genetic ancestry may aid in the diagnosis and treatment of patients with different genetic variants.

LA entails statistically inferring the ancestry of each genomic region of an individual by using dense genotypic data and mapping it to possible contributing ancestral populations. The rapid development in technology and genotyping of patients have enabled scientists to infer ancestry for larger underrepresented cohorts.[27] SNPs are used for ancestry inference and depend on well-representative reference genomes.[42] We can now classify phenotypic variation in such a way that may aid in understanding variations in health outcomes among racial/ethnic groups without the necessity of reporting on self-identified race/ethnicity. There are apparent differences in disease risk between different population groups due to differences in social factors, genetics, environment, lifestyle, comorbidities, and complex interactions between these factors. Genetic ancestry should be incorporated into scientific research to avoid biased assumptions that all population groups have the same disease prevalence. Furthermore, understanding the population differences may aid in the development and enhancement of therapeutic interventions in all populations and consequently provide disparity-reducing benefits for populations with poorer health outcomes.

Recording the patient's self-identified race or ethnicity on the informed consent form of genomic association studies will ultimately become redundant, as the relevant ancestry information (both global and local ancestry) can be accurately inferred with genotypic data from the patient's DNA sample. This information will provide vital clinically relevant information while maintaining privacy and preventing political agendas. Scientists need to reach consensus with regards to the choice of phenotypic descriptors and give clear uniform definitions and descriptions of these terms.

The clinical dilemma of missing traits within complex diseases

Assessment of the clinical relevance of detected gene variants, especially in patients lacking a clear family history, cannot be determined without the availability of well-defined phenotypes. Correct phenotyping remains challenging for many disorders that are complex and multifactorial in nature since the phenotypic presentation of these disorders often differs from person to person. A good example of such a complex disorder with a genetic (i.e., heritable) component is Parkinson's disease (PD). PD is a neurodegenerative disease, characterized by the progressive loss of dopamine-producing neurons in the brain's substantia nigra pars compacta and the formation of proteinaceous inclusions of misfolded α-synuclein, so-called Lewy bodies, in the surviving neurons.[43] In persons with PD the death of these dopaminergic neurons results in several motor (i.e., movement-related) symptoms including rigidity and tremor,[44] which are preceded by a range nonmotor symptoms such as sleep disturbances, hallucinations and loss of smell.[45,46]

PD usually affects those over the age of 65, however, it can present in much younger individuals. Although PD can be reliably diagnosed at autopsy, there is no diagnostically conclusive test for its diagnosis in living individuals. Currently, PD is clinically diagnosed based on the four cardinal symptoms: bradykinesia, rigidity, tremor and postural instability, through the use of diagnostic guidelines such as the UK Parkinson's Disease Society Brain Bank Research criteria.[47] However, these symptoms vary widely between affected individuals and their severity and occurrence change over time. Therefore, the timing of patient screening by a neurologist and the accuracy or inaccuracy of their health records, including that of their family members, can greatly affect the reliability of the diagnosis. Moreover, several of the PD symptoms are shared by a range of atypical parkinsonisms, dementias, and other tremor disorders. This complicates PD diagnosis and accurate phenotyping, and often results in PD misdiagnosis, especially in the early stages of disease.[48]

Brain imaging including magnetic resonance imaging (MRI), single photon emission computed tomography (SPECT) and positron emission tomography (PET) can help increase PD diagnostic accuracy.[49] However, neuroimaging for PD is not yet recommended for routine use in clinical practice. In the developing world, PD diagnosis is even more difficult given that clinical diagnosis requires a neurologist, preferably a movement disorder specialist (i.e., a specialized neurologist), both of which are severely lacking in developing countries.[50] For instance, a recent study reported that 17 of 28 surveyed African countries had less than one available neurologist per million people and that 11 of these countries completely lacked movement disorder specialists.[50]

Due to the nature of the disease, there are also limited samples that can be used to validate a PD diagnosis. In cancers, for instance, biopsies can be taken to understand the nature of a tumor—whether it is benign or malignant.[51] In PD, the affected tissue (i.e., the brain) is not easily accessible, therefore, blood or cerebrospinal fluid samples are used for further testing. However, there are still limited tests that can be conducted on these samples, and efforts to identify and implement reliable PD biomarkers in accessible tissue are still ongoing.[52] The most efficient testing that can be done is on the DNA obtained from the samples. We can use the DNA to screen for known pathogenic variants in known PD-associated genes. Genetic panels are available that usually combine PD genes as well as those of other neurodegenerative disorders, such as Alzheimer's disease.[53–56] These panels can be expensive to run and are limited to the genes that have been conclusively implicated in PD. Therefore, they are generally used for research purposes as opposed to clinical diagnostic testing. Moreover, PD genetic testing is generally only done/offered in a clinical setting for cases where there is a family history of disease. Notably, this type of genetic testing has also currently limited utility in an African context, given that the genetic causes of PD likely differ to those in the rest of the world.[57,58]

Like many complex disorders, PD is not only heterogenous in its clinical presentation but also its underlying biology.[59] More recently, PD is being recognized as mixture of different genetic, clinical and epidemiologically heterogeneous disorders with diverse clinical subtypes and even different phenotypes within each of these subtypes.[60] The implementation of more accurate phenotyping according to these subtypes, may aid the future application of effective genetic testing. However, we are still far away from this, given the added complexity of GEI factors, such as agricultural pesticide exposure in persons with the CYP2D6*4 allele variant, in persons who are at increased risk of PD.[61] Since numerous environmental carcinogens and around 25% of clinically important drugs are metabolized by cytochrome P450 2D6 (CYP2D6), it has been suggested that the same approach should be used for low penetrance variants as applied in high penetrance genetics such as BRCA1/2 previously referred to, while also taking into consideration the effect of environmental exposures.[62]

Autism is another example of a disorder where phenotyping is difficult. It has a significant genetic component and similar to PD, can segregate in families.[63] Autism is a neurodevelopmental disease characterized by language impairments, social and communicative deficits, and repetitive behaviors.[64] Technically, the term "autism" refers to a spectrum of disorders, hence they are characterized as autism spectrum disorders (ASDs). As such, the disorder can vary greatly from one individual to another. ASD is usually diagnosed with the assistance of the Diagnostic and Statistical Manual of Mental Disorders (DSM) written by the American Psychiatric Association (APA),[65] by assessing an individual's development and behavior. However, there is no diagnostically conclusive test available. Furthermore, with ASD and related disorders displaying an overlap in neuropathology,[66] there is also no recognized common neuropathology and no specific cell type or brain region that has been uniquely implicated in ASD. Moreover, the symptoms of ASD mentioned above are very broad and can thus manifest in different ways.[67] For instance, repetitive behaviors can refer to repetitive motor movements and use of objects or it can refer to an inflexible adherence to routines. Social deficits can also be difficult to judge, as failure to initiate or respond to social interactions may be a sign of introversion or shyness, and not necessarily a sign of ASD. Therefore, again, the timing of patient screening and the accuracy of their family history can facilitate differential diagnosis. Since ASD primarily affects children, this can also affect the accuracy of a diagnosis. For example, young children may not be able to explain their symptoms or orphaned children may not be aware of a possible family history.

There has been significant progress in our understanding of the genetic basis of ASD, especially using multiple large-scale association studies.[53,68,69] Such studies have resulted in over 100 ASD genes being associated with the disorder.[70] However, the different phenotypes hamper our ability to conclusively implicate variants found in association studies in ASD development or progression. Another confounding factor in ASD, is the presence of comorbidities. Often, psychiatric symptoms, such as depression, anxiety and attention deficit/hyperactivity disorder (ADHD), can be comorbid in children with ASD.[71,72] This again makes it difficult to conclusively implicate variants in ASD development or progression. Moreover, similar to PD, ASD diagnosis and correct phenotyping is impaired due to a lack of awareness, necessary skilled professionals, and medical facilities, especially in the developing world.[73] In addition, most knowledge we have on ASD is from higher income countries and diagnostic tools were developed for these populations. Due to the social aspects of the disorder, culture can have a significant impact on diagnosis.[73,74] This further complicates phenotyping.

Currently, chromosomal microarrays that detect deleted or duplicated DNA segments are the recommended first-tier genetic test for ASD,[75] while NGS, including genetic panels, may be used as second-tier testing to screen for variants associated with ASD. However, similar to PD, many ASD cases may have a complex genetic etiology with multiple variants cumulatively impacting disease risk.

Challenges in framing the genotype versus phenotype versus environment triad

The challenge to link genotype to phenotype, especially for noncoding regions have led to the development of novel sequence based deep learning methods. Systems level functional approaches are becoming increasingly important in which multiomics data are linked and interconnected to biological networks to assist in predicting the impact of noncoding variants on pathogenicity and impact on disease.[15] This multi-prong approach is not only necessary for improved variant detection and functional validation, but also for more directed clinical application and utilization, particularly within the context of a developing country with limited resources.

The recent publication of a White Paper entitled "A Framework for the Implementation of Genomic Medicine for Public Health in Africa" highlighted the need for the implementation of genomic medicine in Africa, as well as the challenges around the actionability of linkages between genotypes and phenotypes.[76,77] Recent advances in genetics are of little value if they cannot be applied clinically. Pathology-driven phenotype integration with genotype and environmental information not only assists with more accurate disease risk stratification but can also benefit patient-centered care Fig. 7.4. For developing countries to benefit from the health advances genomics can offer, new approaches should be developed for translation of research findings into economically viable platforms. Toward this goal, PSGT is presented as a case study in the framework developed for implementation of genomic medicine in Africa, as applied in South Africa using an integrated service and research approach Table 7.1.[76,77]

Together, phenotype, genotype and EF form the central core of a genetic service platform. Using a multiomic data framework, input from this triad is combined within a well-controlled regulatory

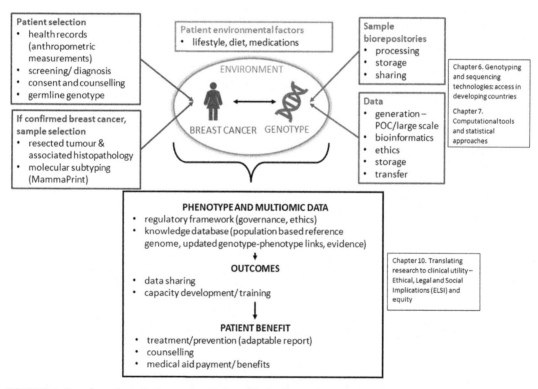

FIGURE 7.4 Overview of a pathology-supported genetic testing systems approach.

How the appropriate selection of phenotype and integration between genotype and environment can benefit patient-centered care in the genomic era. All captured genetic and environmental variables descriptors are used to either explain or predict the phenotype which in turn can be modified to prevent or treat the underlying complex disease.

Images from Word Stock images, open for use.

Table 7.1 Application of using a genetic systemic approach based on risk stratification in noncommunicable diseases.

System approach	Risk stratification
Genetic counseling Clinical risk profile Family history Pathology data	Personal risk Ancestral risk Familial risk Histopathology Treatment failure/side-effects
Define the testing approach for high penetrance genes associated with the pathology	Screen affected member to identify the causative mutation(s) Screening of other family members for the same mutation
Pretest genetic counseling	Test value in family member without pathology Test value in family member with pathology Inheritance risk for offspring
Timing of the genetic test	Depending on the penetrance and associated disease severity
Select variant test option	Mutation-specific Population-specific Full gene screen
Optional research participation for future studies Data storage Data decoding Specimen Destruction Data sharing	Pharmacogenomic treatment modalities of the pathology
Clinical utility	Comprehensive report for clinician and patient integrating: Family history Medical history Risk factors Pathology data Genomic data and Recommendations including: Lifestyle, diet adjustments Change of treatment considerations
Medical aid reimbursement	Private sector
Comprehensive healthcare	State sector

Modified from Kotze MJ, Lückhoff HK, Peeters AV, Baatjes K, Schoeman M, van der Merwe L, Grant KA, Fisher LR, van der Merwe N, Pretorius J, van Velden DP, Myburgh EJ, Pienaar FM, van Rensburg SJ, Yako YY, September AV, Moremi KE, Cronje FJ, Tiffin N, … Schneider JW. Genomic medicine and risk prediction across the disease spectrum. Critical Reviews in Clinical Laboratory Sciences 2015;52(3):120−137. *https://doi.org/10.3109/10408363.2014.997930.*

framework and an updated genetic database Fig. 7.4.[77] Key outcomes for this framework would include the potential for data sharing and research and capacity development within the developing country. The overarching benefit of this being economically viable patient-centered benefit for improved disease screening and prevention as well as treatment stratification.

Proof of principle of viability of the triad within the proposed framework

Despite the limitations of genomic research, accurate phenotyping has resulted in successful implementation of differential disease diagnosis and targeted treatment of complex diseases, including cancer which has been leading the field of precision medicine over the last 2 decades. Breast cancer in particular has been a major driver due to the high prevalence and tumor heterogeneity resulting in diverse clinical outcome.[78,79] Phenotyping of breast cancers are based on the histological subtype, molecular subtype, hormone receptor status by immunohistochemistry, histological grading and tumor staging, including tumor size, lymph node involvement and metastasis.[78,80] In addition to patient factors, these tumor descriptors are used to aid treatment decisions. The addition of molecular subtyping as either luminal A, luminal B, human epidermal growth factor receptor-2 (HER2)-enriched and basal type, have highlighted the importance for the use of genomics to aid in risk stratification and treatment.[81] Use of nonmolecular descriptors alone, may often lead to chemotherapy under-or overtreatment as demonstrated in a double-blind randomized controlled trial,[82] and this knowledge was translated into several multigene-assays for risk assessment in patients with early-stage breast cancer.[83–85]

The advent of gene expression profiling of breast tumors has enabled more accurate risk prediction for relapse and death using gene profiling in addition to germline BRCA1/2 testing and for prevention of secondary cancers.[86,87] For example, MammaPrint, is a 70-gene signature test which can determine the risk of metastasis in early-stage breast cancer with the benefit to reduce chemotherapy overtreatment and was extensively studied in South Africa.[9,82,83] This microarray test, approved by the Food and Drug Agency (FDA) in 2007, has already been successfully implemented within a developing health economic context to assist with the clinical decision-making process in hormone receptor-positive breast cancer.[9,85] While several tests are available for the same clinical indication, only a few have both FDA approval and level 1A evidence of clinical utility.[84] Once diagnosed, estrogen receptor (ER) and progesterone receptor (PR) positive and HER2 negative breast cancer patients with tumors of up to 5 cm and with up to three nodes involved are currently the only group for which MammaPrint is indicated in South Africa Fig. 7.5.[83,85] Exclusion of ER-, PR- and HER2-tumors was based on a critical evaluation of the added cost-risk-benefit the additional genomic information would add to standard clinicopathological management of breast cancer. With implementation of the PSGT approach, preselecting the appropriate target patient population facilitates the correct clinical interpretation.

Based on Fig. 7.4, the interdependence of appropriate phenotype selection integrated with the genotype is highlighted by the PSGT approach to breast cancer risk stratification Fig. 7.6. The initial phenotype may prompt the need for collection of patient-specific information to assist with downstream risk stratification and targeted treatment. Since comorbidities are known to be contributing factors to adverse breast cancer outcome, inclusion of a comprehensive health questionnaire may assist with identifying environmental exposures potentially acting as cofactors.[88,89] Tumor histopathology guides patient selection for further molecular tumor profiling Fig. 7.5. In cases suitable for the

Summary of Results

Clinical Risk Assessment in the MINDACT Trial[2]

ER Status	HER2 Status	Grade	Nodal Status	Tumor Size	Clinical Risk in MINDACT
ER positive	HER2 negative	Well differentiated (Grade 1)	Node-negative	≤ 3cm	Low
				3.1-5cm	High
			1-3 positive nodes	≤ 2cm	Low
				2.1-5cm	High
		Moderately differentiated (Grade 2)	Node-negative	≤ 2cm	Low
				2.1-5cm	High
			1-3 positive nodes	Any Size	High
		Poorly differentiated or undifferentiated (Grade 3)	Node-negative	≤ 1cm	Low
				1.1-5cm	High
			1-3 positive nodes	Any Size	High
	HER2 positive	Well differentiated OR Moderately differentiated (Grade 1 / Grade 2)	Node-negative	≤ 2cm	Low
				2.1-5cm	High
			1-3 positive nodes	Any Size	High
		Poorly differentiated or undifferentiated (Grade 3)	Node-negative	≤ 1cm	Low
				1.1-5cm	High
			1-3 positive nodes	Any Size	High
ER negative	HER2 negative	Well differentiated (Grade 1)	Node-negative	≤ 2cm	Low
				2.1-5cm	High
			1-3 positive nodes	Any Size	High
		Moderately differentiated OR Poorly differentiated or undifferentiated (Grade 2 / Grade 3)	Node-negative	≤ 1cm	Low
				1.1-5cm	High
			1-3 positive nodes	Any Size	High
	HER2 positive	Well differentiated OR Moderately differentiated (Grade 1 / Grade 2)	Node-negative	≤ 1cm	Low
				1.1-5cm	High
			1-3 positive nodes	Any Size	High
		Poorly differentiated or undifferentiated (Grade 3)	Any	Any Size	High

[1] Buyse, et al. J Natl Cancer Inst. 2006 Sep 6.98(17): 1183-92.
[2] Cardoso, F et al. N. Engl J Med. 2016 Aug 25. 375 (8): 717-29.

FIGURE 7.5 Clinical risk assessment as defined in the MINDACT trial.

The modified adjuvant online clinical risk assessment tool was used in the MINDACT (Microarray In Node negative Disease may Avoid Chemotherapy) trial to evaluate the high risk clinico-pathological characteristics

MammaPrint service, the test result reports a high- or low risk based on the MammaPrint Index (MPI) with a range of $+1$ to -1 (high risk: 1 to 0; low risk: $+0.001-1$). A high-risk recurrence score Fig. 7.7 indicates the need for chemotherapy treatment to reduce the risk of metastasis. Whereas a low recurrence risk score Fig. 7.8 is associated with no additional benefit of chemotherapy when added to endocrine therapy. In the Stockholm tamoxifen (STO-3) trial, 1976 to 1990, of 652 postmenopausal breast cancer patients, approximately 25% of the low-risk category with an MPI $> +0.355$ is reported as ultra-low risk of recurrence, representing 15% of the study population.[90]

These are patients with a very low Risk-of-Recurrence up to 20-year with limited or no hormone therapy. Since the MammaPrint test cannot predict the risk of secondary cancers, further testing for BRCA1/2 variants is warranted even in hormone receptor positive cancers generally considered to be low risk for carrying pathogenic germline variants compared to triple negative breast cancer.[87] Previously, age of onset as well as a strong family history of breast cancer prompted the need for germline genetic testing of BRCA1/2 variants. However, interaction between gene variants and lifestyle factors contribute to variable expression in both hereditary (familial) and nonhereditary (polygenic) breast cancer pointing to an increased need for genetic screening in healthy females, patients with sporadic cancer as well as familial breast cancer patients.[91]

Risk-benefit analysis strongly supports the early use of POC genetic variant screening in breast cancer management, particularly within multi-ethnic communities such as South Africa with a high founder mutation rate.[87] Integration of germline and tumor genetics not only facilitate the determination of familial and endocrine therapy associated risk but may also provide insight into the prevention of other secondary cancers due to the pleiotropic effect of some variants.[86,87] In addition, it enables association with potential drug responses as well as common epigenetic processes overlapping the expression of noncommunicable diseases.[9,91] Addressing risk factors such as estrogen exposure, diet, and lifestyle is key in the prevention of polygenic sporadic breast cancer due to low penetrance gene variants.[91] Early involvement of genetic counseling assists with psychological support, wellbeing of patients, as well as guidance with the implications of genetic testing for self and others, such as close family members.

Successful translation of genomics into clinical practice requires proof of cost-effectiveness in a target group most likely to benefit from gene-based intervention, which in turn is essential for development of policy-making frameworks.[92] When implementing genomic medicine into a developing country framework this is key to consider. To mitigate concerns of inappropriate test ordering and doctors' not using genomic test results in treatment decisions, authorization policies may be developed by medical insurers, each with their own process in place for obtaining informed consent. The first example of a health technology assessment (HTA) performed for this purpose in South Africa, led to reimbursement of the above-mentioned MammaPrint service based on the outcome of a pilot

in patients with breast cancer, to safely reduce chemotherapy use based on a low risk genomic profile. Clinical assessment is based on the estrogen receptor (ER) and HER2 tumor status, the tumor grade, the nodal status and tumor size. Based on this, early-stage breast cancer patients are then clinically defined as low or high risk. This tool is used to select breast cancer patients for who MammaPrint is currently indicated worldwide.

From Cardoso F, van't Veer LJ., Bogaerts J, Slaets L, Viale G, Delaloge S, Pierga JY, Brain E, Causeret S, DeLorenzi M, Glas AM, Golfinopoulos V, Goulioti T, Knox S, Matos E, Meulemans B, Neijenhuis PA, Nitz U, Passalacqua R, ... MINDACT Investigators. 70-Gene Signature as an Aid to Treatment Decisions in Early-Stage Breast Cancer. The New England Journal of Medicine 2016; 375(8):717–729. https://doi.org/10.1056/NEJMoa1602253.

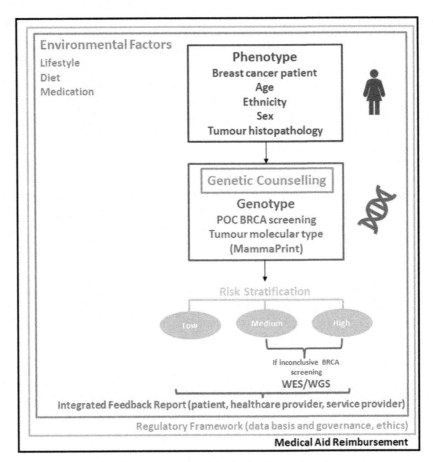

FIGURE 7.6 An integrated platform for breast cancer risk stratification and management.

Proof of concept of how integrated pathology driven care like the use of MammaPrint for molecular phenotyping and germline screening can be combined to offer personalized, yet cost-effective treatment and prevention strategies.

Images from Word Stock images, open for use.

program and database audit.[83] The MammaPrint prescreen algorithm developed as a cost-minimization strategy in this study was supported and further refined based on the level 1A evidence for clinical utility provided by the prospective double-blind MINDACT (Microarray In Node negative Disease may Avoid Chemotherapy) trial.[82] The finding that the same tumor in the same patient may result in differences in risk scoring depending on the test used, furthermore led to the PROMIS (Prospective Study of MammaPrint in Breast Cancer Patients With an Intermediate Recurrence Score) trial to address the question: "whether to print or type".[93] Such ethical dilemmas can also be resolved by long-term follow-up studies and database curation linked to surveillance programs integrating germline and tumor genetics.

Summary of Results: **HIGH RISK**

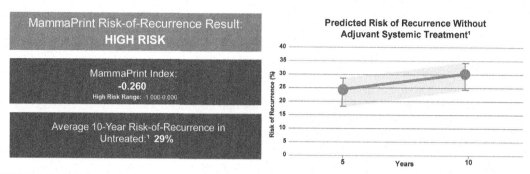

FIGURE 7.7 High risk-of-recurrence MammaPrint result.

Example of the results obtained in a patient with early-stage breast cancer found to have a high risk of distant metastasis based on a MammaPrint Index inferior or equal to 0.

Personal communication courtesy of Director Marketing and EU Medical team Agendia. Cardoso F, van't Veer LJ, Bogaerts J, et al. 70-Gene signature as an aid to treatment decisions in early-stage breast cancer. N Engl J Med. 2016;375(8):717–729. https:// doi.org/10.1056/NEJMoa1602253

Summary of Results: **LOW RISK**

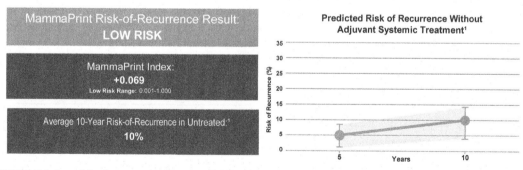

FIGURE 7.8 Low risk-of-recurrence MammaPrint result.

Example of the results obtained in a patient with early-stage breast cancer found to have a low risk for metastasis based on a MammaPrint Index of more than 0 and up to 1. The average 10-year risk-of-recurrence in untreated persons is 10%.

Personal communication courtesy of Director Marketing and EU Medical team Agendia. Cardoso F, van't Veer LJ, Bogaerts J, et al. 70-Gene signature as an aid to treatment decisions in early-stage breast cancer. N Engl J Med. 2016;375(8):717–729. https:// doi.org/10.1056/NEJMoa1602253

Key to this aim of medical insurance coverage of genomic testing is the establishment of secure databases capturing the patient phenotypic information including family and medical history, pathology results if available and environmental factors and then integrating this information with germline and when appropriate tumor molecular subtyping, to produce an integrated report for the patient and healthcare provider. This should be provided within a secure framework of ethical and data protection policies and high-throughput data generation, analysis, and storage facilities.[77] Continual identification of new variants necessitates a research arm within this framework to allow for the identification and inclusion of newly identified novel variants, especially within diverse populations.[94] Establishing validated polygenic risk scores provides opportunities for capacity development and training within resource limited settings, as well as ensuring the use of the most up to date variants and comprehensive technologies inclusive of all ancestries.

Development of appropriate ethical and legal frameworks are essential for studies involving return of research results to individual study participants, especially when new insights gained from the extended study cohort are considered for inclusion in the feedback report using a three-prong approach focused on inherited, lifestyle-triggered and therapy-induced risk stratification.[40] An effective genomic medicine model requires a research translation pipeline including skills development across different healthcare disciplines, as well as community engagement and public education. In this context, PSGT strives to incorporate clinically relevant research findings into an existing body of knowledge for disease prevention, prognosis and/or targeted treatment. Specialized training in data management for report generation is currently provided through the Open Genome Project (https://www.gknowmix.org/opengenome/), which aims to combine POC genomics with genetic counseling to increase access to personalized medicine. The infrastructure to perform PSGT of multiple NCD pathways for assessment of different aspects of a disease in the same patient, was put in place through government investment in information technology over more than a decade. Ultimately, first-tier POC platforms for sample processing and accurate data capture will be interconnected with high-throughput technologies and sample storage facilities for computational data analysis. Biochemistry tests, immunohistochemistry and other health assessments with the potential to uncover genetic underpinnings of complex multifactorial diseases, enables the establishment of genomic databases in parallel to routine patient care https://cansa.org.za/establishing-cancer-genomics-registry-to-support-implementation-of-personalised-medicine/.

Pathology-supported genomics: Report overview
Background introduction

This case report highlights how using genome scale NGS technologies such as WES can add clinical value after the most common or major single disease-causing variants were excluded. By using a multiprong PSGT approach and combining patient personal and family history, lifestyle and nutritional data, as well as biochemistry with results of low-penetrance gene variant screening, the importance of lifestyle adjustment and prevention strategies is highlighted for future clinical management.

Pulmonary embolism (PE) diagnosed in this patient, is one of the highest causes of cardiovascular related deaths worldwide.[95] Thromboembolism is mainly due to contributing risk factors affecting either[1] endothelial blood vessel damage,[2] hypercoagulation and[3] blood flow irregularities. These three components are known as Virchow's triad. The most common genetic variants associated with thromboembolism include deficiencies of antithrombin III (AT-III), protein C (PC), protein S (PS) and well-established single nucleotide variants (SNVs) in the prothrombin factor II (20210 G > A, prothrombin) and factor V (1691 G > A, Leiden) genes, as well as multiple new variants identified in GWAS.[96–98] An underlying genetic contribution or cause for PE should be taken into account in females with recurrent miscarriages or stillbirths, any person with a family history and no other risk factors, recurrent PE, or unexplained thromboembolic events between 45–50 years of age.[97] In persons diagnosed with PE, the gold standard for prevention has been vitamin K antagonists such as warfarin which aim to maintain an international normalized ratio (INR) value between 2.0 and 3.0.[99,100] The European Respiratory Society recognizes the value of pharmacogenetic testing to improve anticoagulation control.[100] Variants of CYP2C9, CYP4F2, VKORC1 and the CYP2C cluster are the most common underlying genetic causes for differences in warfarin metabolism.[101] In addition, multiple nongenetic causes may also lead to poor INR control such as poor patient drug compliance, drug-drug interactions, drug absorption and clearance issues and high dietary vitamin K intake and should also be considered before investigating genetic causes only.[101,102] Cases of warfarin resistance require further examination to understand the underlying driving factors.[102]

Case presentation

A female patient was referred for comprehensive genomic screening due to a persistent inability to increase the INR levels to the therapeutic range on warfarin treatment. This adult patient was initiated on warfarin treatment after presenting with a PE at a relatively young age. Her medication included the use of carvedilol, losartan, doxazosin, amlodipine and allopurinol. She was not on any hormone replacement therapy. The only possible drug which could interact with warfarin was allopurinol. This potential interaction is listed as moderate and usually tends to increase INR levels. The patient was compliant with her medication use with no documented increased consumption of vitamin K. Exclusion of occult cancer or acquired causes influencing warfarin absorption, metabolism, and clearance was also an important consideration. She had recurrent pregnancy losses, and a known family history of PE as her father also developed this at an unknown age. Laboratory screening for antiphospholipid syndrome, antithrombin III deficiency, and activated protein C resistance were negative.

Questionnaire based assessment of clinical characteristics to guide genetic testing and interpretation

Prior to genetic testing, clinical assessment showed a raised body mass index (BMI) of 29.62 kg/m^2, with a high blood pressure of 160/90 mmHg Fig. 7.9. Her chemical pathology results revealed iron deficiency with low serum iron (7 umol/L), transferrin (2.62 g/L) and transferrin saturation (11%), with normal serum ferritin (73 ng/mL) levels. In addition, she had a raised C-reactive protein count (44 mg/L) and fasting glucose with a borderline high normal HbA1c (5.9%). Her lipid levels were not provided.

Her lifestyle assessment revealed that she was a nonsmoker, did not consume alcohol and maintained a low physical activity status. Nutritionally, her intake of saturated and trans fats but also of fruit, vegetables, fiber and folate in general were limited, and she took regular vitamin B10 supplementation.[9] A matrix for data integration and semiautomated report generation using PSGT was used

FAMILY MEDICAL CONDITIONS	AGE OF ONSET	RELATIONSHIP
Pulmonary Embolism		Father
PERSONAL MEDICAL HISTORY	**AGE OF ONSET**	**MEDICATION / OTHER**
Cardiomyopathy	58 years	Carvedilol; Losartan
Hypertension	44 years	Doxazocin; Amlodipine
Pulmonary Embolism	37 years	Warfarin
Recurrent Pregnancy Loss		
Stroke	62 years	
Chronic kidney disease stage IV	50 years	Allopurinol
Medication side effects/failure		Warfarin resistance **
CLINICAL ASSESSMENT & PATHOLOGY TESTS	**VALUE**	**EVALUATION VALUES**
C-reactive protein (CRP)	**44 mg/L - High**	<5 mg/L
Glucose, fasting	**6.8 umol/dL - High**	3.3-5.5 umol/dL
HbA1c	5.9 %	4-6 %
Serum ferritin	73 ng/mL	15-300 ng/mL
Serum Iron	**7 umol/L - Low**	12.5-30 umol/L
Transferrin	2.62 g/L	2-3.6 g/L
Transferrin saturation	**11 % - Low**	20-50 %
Blood pressure – Systolic	**160 mmHg - High**	<140 mmHg
Blood pressure – Diastolic	**90 mmHg - High**	<90 mmHg
Weight	73 kg	
Height	1.57 m	
Body mass index (BMI) Adult	**29.62 kg/m^2 - High**	18.5-24.9 kg/m2
Contraceptive pill	No	
Hormone replacement therapy	No	
LIFESTYLE ASSESSMENT	**SCORE**	
Physical activity	**Low**	
Smoker	No	
Alcohol consumption	None	
NUTRITION ASSESSMENT	**SCORE**	
Fat intake, saturated & *trans* fats	19 - Low (excellent)	
Fruit, vegetables, fibre intake	**4 - Very Low**	
Folate intake	**10 – Low *****	
Food supplements taken daily	vitamin B10	

FIGURE 7.9 Summary of patient clinical information as extracted from a pathology supported genetic testing report.

An extract from the patient report showing family and personal history, clinical assessment and provided pathology test results, lifestyle and nutrition assessment.

With permission by author MJ Kotze.

to identify risk factors underlying treatment response and disease pathogenesis, as previously described.[9] This approach enables the contextualization of abnormal pathology or medication side-effects/failure as a consequence of genetic or environmental risk factors, or both.[9]

The inability to control her INR levels on warfarin, together with her family history of PE and cardiovascular conditions and risk factors, prompted holistic genetic investigation to identify an underlying causative link for her conditions. Given the patient's early onset of PE together with a personal history of recurrent pregnancy loss and a family history of PE, a possible underlying genetic cause for this was also investigated. In addition to her history of PE, she has other established cardiovascular conditions and complications including hypertension, cardiomyopathy, stage IV chronic kidney disease and stroke. Her raised BMI and physical inactivity added to her risk for recurrent PE and other cardiovascular complications.

Genetic approach

Exclusion of common variants relevant to the clinical profile or insufficient to explain the pathology, prompted extended genetic testing using WES. Using the PSGT approach algorithm including multiple pathways (1−6 as indicated below) as a first-tier test for risk stratification, the personal and family history, questionnaire-based lifestyle assessment, medication use/reported side effects and previous pathology results were integrated to guide selection of key biochemical pathways for WES screening.[103]

1 Drug metabolism (warfarin)
2 Lipid and lipoprotein metabolism (including APOE-vitamin K-cholesterol Pathway)
3 Homocysteine and folate metabolism (MTHFR-folate-homocysteine Pathway)
4 Hemostasis and thrombophilia (FII/FV-blood clotting Pathway)
5 Iron metabolism (deficiency/overload) (HFE/TMPRSS6 iron overload/deficiency Pathways)
6 Inflammation (metabolic syndrome features)

While the report focused on key pathways relevant to the patient risk profile, i.e., the warfarin drug metabolism, hemostasis and thrombophilia, the first-tier genotype also served as an internal control to prevent sample/data mix-up during WES/WGS usually sourced to a high-throughput reference laboratory.

WES methodology

WES was performed on the Ion Proton apparatus (Central Analytical Facility, Stellenbosch) using the method described by Van der Merwe et al. (2017).[62] A rapid POC test kit that includes intra-genic/ promoter (Table 7.2) gene variants were incorporated into the pathology-supported genetic testing framework applied for WES analysis of specific clinically relevant gene panels (A-C described below) partly informed by the results of the above-mentioned six pathways.

Table 7.2 Variants detected in the WES screen.

Gene test	Variant detected	Classification and interpretation
VKORC1	rs9923231: 1639 GG rs7200749 (p.P83L): c.248 CT	Pharmacogenetics variant: Increase warfarin dose
CYP4F2	rs2108622 (p.V433M): 297 GA	Pharmacogenetics variant: Increase warfarin dose

Gene panel A. Pharmacogenetics genes (warfarin): VKORC1, CYP2C9, CYP4F2, CYP2A6, CYP1A2, GGCX, CYP3A4, CYP2C8, CALU, CYP1A1, CYP2C19 Fig. 7.10.

Gene panel B. Inherited thrombophilia genes: FII, FVIII, FV, FXII, PROC, PROS1, SERPIND1 Fig. 7.10.

Gene panel C. Cancer susceptibility genes: BRCA1, BRCA2, ATM, CHEK2, CDH1 PTEN PALB2, STK11, TP53, BARD1, MSH6, RAD51B, BRIP1, EPCAM, NBN, RAD51C, FANCC, NF1, RAD51D, MLH1, PMS1, RECQL, MRE11A, PMS2, MUTYH, MSH2, RAD50, XRCC2.

Screening tests for cancer performed as part of the clinical work-up for thrombosis in the patient, supported additional WES data analysis given the knowledge that cancer cells may release chemicals that stimulate the body to produce more clots.

The Ion AmpliSeq Exome RDY kit was used for library construction and samples were sequenced with the One Touch workflow. Variant calling was performed on the Torrent Server against an ethnically concordant major allele reference genome (based on GRCh37.p13, Annotation Release 105). Variants were annotated by wANNOVAR (http://wannovar.wglab.org/) and prioritized against a primary list of -related genes listed in Fig. 7.10. The mean exome-wide depth of coverage was at least 120× and uniformity was >92%. The CNVkit tool (https://github.com/etal/cnvkit) was used to detect copy number variants.[104]

Bioinformatics tools

Variant classification into pathogenic, likely pathogenic, variant of uncertain significance (VUS), likely benign, or benign is based on the American College of Medical Genetics and Genomics (ACMG) guidelines for diseases with Mendelian inheritance. Reclassification of a VUS is expected as new research findings become available; a downgrade from pathogenic to benign could have

Clinically relevant genes screened with WES			
CYP2C9 NM_000771.4	CYP2A6 NM_000762.6	CYP1A2 NM_000761.5	GGCX NM_000821.7
VKORC1 NM_024006.6	CYP3A4 NM_017460.6	CYP2C8 NM_000770.3	CALU NM_001219.5
CYP4F2 NM_001082.5	CYP1A1 NM_000499.5	CYP2C19 NM_000769.4	SERPINC1 NM_000488.4
EPHX1 NM_000120.4	CYP2C18 NM_000772.3	CYP3A5 NM_000777.5	FV NM_000130.5
FII NM_000506.5	PROC NM_001375607.1	SERPIND1 NM_000185.4	PROS1 NM_001314077.2
FVIII NM_000132.4	FXII NM_021489.3		
		*Shaded blocks relates to genes associated with inherited thrombophilia	

FIGURE 7.10 Clinically relevant genes screened with WES.

Genes included in a virtual panel screened for warfarin metabolism and inherited thrombophilia, using WES.

With permission by author MJ Kotze.

significant implications. The ClinVar database (https://www.ncbi.nlm.nih.gov/clinvar) and the score provided by the VarSome calculator (https://varsome.com) were used for interpretation. WES data may be re-analyzed in the future, and a new report issued, as potentially novel genes become available and bioinformatics tools improve.

Results

Extended pharmacogenetic WES analysis of panel A added significant clinical value, as evidenced by detection of CYP4F2 rs2108622 (1297G > A, V433M) associated with an increased warfarin dosage requirement. In addition to VKORC1 rs9923231 (−1639 GG), rs7200749 (c.248 C > T, p.P83L) was detected in the same gene associated with warfarin resistance in this patient Table 7.2. Genotype-guided warfarin dosing using the calculator available at www.warfarindosing.org was completed for the patient using the results in the genetics and clinical characteristics (Table 7.2, Figs. 7.9 and 7.10). WES from panel B did not identify a single pathogenic, causative gene variant associated with inherited thrombophilia. The cancer susceptibility gene screen (panel C) identified a susceptibility to cancer, which would require confirmation in the laboratory using gold standard Sanger sequencing upon the request of the treating clinicians.

The use of quality indicators comparing genotyping results of overlapping DNA sequences excluded the possibility of sample and/or data mix-up during extension of POC testing to NGS. Screening of the APOE-vitamin K-cholesterol Pathway detected the apolipoprotein E (APOE) E4 allele, of which the clinical relevance was determined by using a questionnaire-based approach for assessment of lifestyle risk factors to facilitate clinical interpretation of this context-dependent SNV.[105] Dietary vitamin K deficiency is associated with bleeding and response to Warfarin (vitamin K anticoagulant), which may be influenced by dietary factors and liver function. Hepatic uptake of chylomicrons, and vitamin K carried by chylomicrons, depends largely on the APOE genotype-phenotype relationship. Genetic variation was also detected in the methylenetetrahydrofolate reductase (MTHFR) gene, which reinforces the importance of folate intake above the recommended daily dose to prevent or restore dysfunction of the methylation pathway when clinically indicated.[106] Heterozygotes with one copy of the 677C > T mutation only have 65% of normal MTHFR enzyme activity, which is associated with an approximate 20% increase in homocysteine levels when folate intake is low. No other variants were detected in the other pathways.

Recommendations from the results

Detection of the VKORC1 -1639 GG genotype (Warfarin insensitive) with the clinically validated warfarin pharmacogenetics POC assay that included three SNVs in the VKORC1 and CYP2C9 genes, is compatible with a Warfarin resistant phenotype. However, this finding does not fully explain the lack of INR increase in the case studied, even on high Warfarin doses of 35 mg/day. Calculation of the warfarin dose using the tool available at www.warfarindosing.org by taking into account genetic variation, INR, medication use, as well as the lifestyle data provided in table 1 of this case report, resulted in a Warfarin dosage prediction of 7.2 mg/day. Since this is much less than the use of 35 mg/day at the time of testing, it is possible that both the VKORC1 and CYP4F2 variants, as well as APOE e4 heterozygosity (not captured in the online tool used) also contribute to warfarin resistance in this patient.

A single gene underlying thrombophilia was not detected by WES, and although a monogenic cause cannot be completely excluded, a cumulative effect is more likely. In addition, the patient's modifiable lifestyle risk factors such as raised BMI and relatively low physical activity may also play a significant role. Warfarin resistance or inefficacy includes thorough clinical investigation of underlying risk factors for thrombosis, including occult cancer. Health recommendations provided, were based on lifestyle factors that may trigger the effect of low-penetrance gene variants associated with the metabolic syndrome, including the e4 allele detected in the APOE gene. This multi-functional gene is involved in cellular uptake of lipoproteins which transport fat-soluble nutrients such as vitamin K involved in blood clotting. Genotype-phenotype correlation highlighted borderline obesity and genetic variation in the MTHFR-homocysteine pathway as clinically relevant targets for risk reduction intervention. Interpretative commenting on the PSGT performed in the patient is provided to the treating clinician for follow-up and health monitoring.

In conclusion, the case report serves as an example of how a pathology-supported testing approach can be applied to identifying genetic risk factors associated with NCDs.[107] The inability to identify high-to-moderate penetrance disease causing variants that may be shared among family members with sharing similar pathology, highlights the necessity to use a multiprong approach for NCDs with a genetic component and to use NGS platforms like WES for extended screening and to assist in formulating a holistic treatment approach. Low-penetrance genes are especially dependent on gene-gene and GEI, and therefor data integration of identified variants with known environment risk factors that may underlie abnormal biochemistry is important.

Notes on patient consent and clinical reporting

Informed consent was obtained by the treating clinician, for application of personalized medicine using an integrated service and research approach (Ethics Reference No: N09/08/224). Additional approval for the WES report following first tier POC testing incorporated into the PSGT platform was provided without modification by the Health and Research Ethics Committee of Stellenbosch University (Ethics Reference No: C21/03/008). All personal identifiers have been removed from figures as well as the written document. Stellenbosch University's Clinical Pharmacologist Prof Eric DeCloedt, Biochemist Prof Susan van Rensburg, genetic counselor Dr. Nicole van der Merwe, registered medical scientist Dr. Armand Peeters, and intern medical scientists involved in return of research results for this case at the time, Dr. Stanton Hector and Ms Kelebogile Moremi, authorized the WES report issued by Gknowmix (Pty) Ltd. as reflected on herein. Consulting registered dietitian Ms Lindiwe Whati is acknowledged for codevelopment of the report template including diet and lifestyle recommendations using the PSGT platform. All personal identifiers of the patient have been removed from figures as well as the written document.

Conclusion

Amidst strong sentiments against using "Biological Race" to stratify patients in providing specific healthcare services, Medical Genetics is one discipline where diagnosis and therapies are considered to be inherently reliant on biological racial identity of a person. In the past decade, personalized medicine has been at the forefront offocusing on the differences in genetic make-up of the individual patient. In some instances, these variances are related to ancestral lineage which can be interpreted as "Race". In settings with limited resources, testing is expensive for diagnosing genetic conditions, and self-reported race is used as a means to explore the possible mutations which are more common in some races than others due to a founder effect or population substructure. However, race should not be used as a "short cut" for genetic testing, and/or as a proxy for risky behavior such as poor compliance or lifestyle. This is unethical. Genetic variations are common not only across races but also within the same race, some even in one family. Thus, personalized medicine is not race-based medicine but rather a genetically stratified medicine for each person. Use of race as a risk factor for certain diseases may give rise to unflattering medical stereotyping. The same applies to family history in individual cases where it can be almost as harmful as an overreliance on race and ethnicity. Assuming that all family members will have the same environmental exposures or genetic susceptibilities to disease may represent a health threat to the individual, in whom both could be involved to a certain degree. We need to promote personalized medicine and not racialized medicine. Efforts should be rather driving in identifying the link between specific genetic variations, how environmental factors influence them, whether these produce dysfunctional proteins and cause blood biochemical abnormalities as an intermediate phenotype predisposing to the particular disease. Clinical Pathology is ideally suited to bridge these knowledge gaps with good examples in applications of genetic testing in patients with cancer and other NCDs.

The 70-gene MammaPrint assay is an excellent proof of principle how the appropriate identification and integration of transcriptional gene profiling versus phenotype versus environment triad can be both economically viable and improve patient morbidity. Merging tumor and germline genetics and biochemical pathways allows for further integration of key modalities involved in disease management such as prevention (breast cancer screening and genetic screening of at-risk persons), treatment (pharmacogenomic testing) e.g., testing for CYP2D6 for underlying endocrine therapy risk, as well as familial versus lifestyle triggered risk.[86] Factors such as an increased BMI in postmenopausal woman play a dominant role in increasing risk for the development of breast cancer in at risk persons.[108,109] In addition, postmenopausal females are at increased risk of cardiovascular events and therefore a holistic and comprehensive screening approach is vital in breast cancer patients.[9,110] Elucidation of the epigenetic factors (including environmental) which contribute to disease variant expression risk, and which can be modulated by early identification and prevention strategies should form a key role in disease prevention and control.[9,88] Comprehensive history taking is required to capture environmental factors which may modify gene expression. This would be targeted for the specific phenotype in question. Examples of this include a lifestyle assessment, physical activity report, smoking status, and alcohol consumption. Approaches with regards to dietary assessment would include formulating a dietary score as determined by saturated and unsaturated fat intake, consumption of fruit, vegetables, fiber and animal proteins.

Most developing countries are characterized by unique genotypic diversity which is still underrepresented in most genetic studies. Despite this lack of representation, great strides are being made to address this. In addition, strategies to implement precision medicine within a developing economic landscape are evolving.[76] There is a pressing need for research capacity building in developing countries and data sharing across health disciplines. Multiple barriers hinder the implementation of personalized genomic medicine on a global scale. Of particular concern is the underutilization of clinical pathology as an intermediate phenotype to identify cumulative disease risk caused by a combination of genetic, lifestyle and other environmental factors, which in turn could be targeted to monitor response to treatment. As discussed in this chapter, critical to this implementation is an appropriate understanding of the genotype-phenotype-environment interaction and how PSGT, aims to bring genomics into the treatment domain for clinical decision-making as an economically viable approach for genomic care in developing countries.

Disclosure

M.J.K. is a nonexecutive director and shareholder of Gknowmix (Pty) Ltd., which has developed a database tool for translation of research into clinical practice.

Glossary

Admixture When two previously isolated populations interbreed resulting in a mosaic of distinct ancestry segments.

Allele Each gene has two copies. An alternative form of a gene on a specific location in a chromosome is known as an allele. Interbreeding between two previously isolated populations resulting in a mosaic of ancestral chromosomal segments.

Ancestry Refers to the genetic ancestry inferred at each genomic region along the genome using the DNA of the individual.

Ethnicity A social construct when people self-identify as a distinct group based on their traditions, cultural beliefs, and lifestyle.

Genetic variation Describes the differences between the DNA sequence of individual genomes. There are two classes of variants i.e., constitutional (germline) variants which are inherited and present in the genome at birth; and postzygotic (somatic) variants which occur after birth and throughout an individual's lifetime.

Genotype The total DNA in all chromosomes of a cell. It is also the genetic contribution to a phenotype and can refer to a specific set of alleles which is inherited at a particular locus.

Phenotype A measurable and sometimes observable characteristic of a person such as eye color, height, disease pathology, etc.

Race A social construct which characterizes people into distinct groups based on their outward appearance (skin color, hair, eyes, etc.)

Trait Distinguishing characteristics which determine a person's appearance.

List of abbreviation

25(OH)D	25-Hydroxyvitamin D₃
ACMG	American College of Medical Genetics and Genomics
ADHD	Attention deficit/hyperactivity disorder
APA	American Psychiatric Association
APOE	Apolipoprotein E
ASDs	Autism spectrum disorders
AT-III	Antithrombin III
BMD	Bone mineral density
BMI	Body mass index
CRP	C-reactive protein
CVD	Cardiovascular diseases
CYP2D6	Cytochrome P450 2D6

DNA	Deoxyribonucleic acid
DSM	Diagnostic and Statistical Manual of Mental Disorders
DZ	Dizygotic twins
EF	Environmental factors
ER	Estrogen receptor
FDA	Food and Drug Agency
FII	Factor II
FV	Factor V
GEI	Genotype-environment interactions
GWAS	Genome-wide association studies
HER2	Human epidermal growth factor receptor-2
HTA	Health technology assessment
INR	International normalized ratio
LDL	Low-density lipoprotein
MINDACT	Microarray In Node negative Disease may Avoid Chemotherapy
MPI	MammaPrint Index
MRI	Magnetic resonance imaging
MTHFR	Methylenetetrahydrofolate reductase
MZ	Monozygotic twins
NCDs	Noncommunicable diseases
NGS	Next generation sequencing
PC	Protein C
PD	Parkinson's disease
PE	Pulmonary embolism
PET	Positron emission tomography
POC	Point-of-care
PR	Progesterone receptor
PROMIS Score	Prospective Study of MammaPrint in Breast Cancer Patients With an Intermediate Recurrence
PS	Protein S
PSGT	Pathology-supported genetic testing
QTL	Quantitative trait loci
SASHG	Southern African Society for Human Genetics
SCT	Sickle cell trait
SNP	Single nucleotide polymorphisms
SPECT	Single photon emission computed tomography
VDR	Vitamin D receptor
VUS	Variant of uncertain significance
WES	Whole exome sequencing

References

1. The International HapMap Project. *Nature*. 2003;426(6968). https://doi.org/10.1038/nature02168.
2. Hallauer AR. History, contribution, and future of quantitative genetics in plant breeding: lessons from maize. *Crop Sci*. 2007;47(S3):S-4—S–19. https://doi.org/10.2135/cropsci2007.04.0002IPBS.
3. Mackay TFC, Stone EA, Ayroles JF. The genetics of quantitative traits: challenges and prospects. *Nat Rev Genet*. 2009;10(8):565—577. https://doi.org/10.1038/nrg2612.

4. Newman H, Freeman F, Holzinger K. Twins: a study of heredity and environment. *Am J Psychiatr.* 1937; 94(1):230−231. https://doi.org/10.1176/ajp.94.1.230 (Chicago: University of Chicago press, 1937.).

5. Hwang L-D, Mitchell BL, Medland SE, Martin NG, Neale MC, Evans DM. The augmented classical twin design: incorporating genome-wide identity by descent sharing into twin studies in order to model violations of the equal environments assumption. *Behav Genet.* 2021;51(3):223−236. https://doi.org/10.1007/s10519-021-10044-0.

6. Fraga MF, Ballestar E, Paz MF, et al. Epigenetic differences arise during the lifetime of monozygotic twins. *Proc Natl Acad Sci U S A.* 2005;102(30):10604−10609. https://doi.org/10.1073/pnas.0500398102.

7. Hirsch J. Behavior genetics and individuality understood. *Science.* 1963;142(3598):1436−1442. https://doi.org/10.1126/science.142.3598.1436.

8. Tukker AM, Royal CD, Bowman AB, McAllister KA. The impact of environmental factors on monogenic mendelian diseases. *Toxicol Sci.* 2021;181(1):3−12. https://doi.org/10.1093/toxsci/kfab022.

9. Kotze M. Application of advanced molecular technology in the diagnosis and management of genetic disorders in South Africa. *South Afr Med J.* 2016;106(6 Suppl 1):S114−S118. https://doi.org/10.7196/SAMJ.2016.v106i6.11012.

10. de Castro MV, Silva MVR, Soares FB, et al. Follow-up of young adult monozygotic twins after simultaneous critical coronavirus disease 2019: a case report. *Front Med.* 2022;9. https://doi.org/10.3389/fmed.2022.1008585, 1008585.

11. Pretorius E, Venter C, Laubscher GJ, et al. Prevalence of symptoms, comorbidities, fibrin amyloid microclots and platelet pathology in individuals with long COVID/post-acute sequelae of COVID-19 (PASC). *Cardiovasc Diabetol.* 2022;21(1):148. https://doi.org/10.1186/s12933-022-01579-5.

12. Hartiala JA, Hilser JR, Biswas S, Lusis AJ, Allayee H. Gene-environment interactions for cardiovascular disease. *Curr Atherosclerosis Rep.* 2021;23(12):75. https://doi.org/10.1007/s11883-021-00974-9.

13. Koyama S, Ito K, Terao C, et al. Population-specific and trans-ancestry genome-wide analyses identify distinct and shared genetic risk loci for coronary artery disease. *Nat Genet.* 2020;52(11):1169−1177. https://doi.org/10.1038/s41588-020-0705-3.

14. Wang H, Zhang F, Zeng J, et al. Genotype-by-environment interactions inferred from genetic effects on phenotypic variability in the UK Biobank. *Sci Adv.* 2019;5(8). https://doi.org/10.1126/sciadv.aaw3538. eaaw3538.

15. Wong AK, Sealfon RSG, Theesfeld CL, Troyanskaya OG. Decoding disease: from genomes to networks to phenotypes. *Nat Rev Genet.* 2021;22(12):774−790. https://doi.org/10.1038/s41576-021-00389-x.

16. Baye TM, Abebe T, Wilke RA. Genotype−environment interactions and their translational implications. *Pers Med.* 2011;8(1):59−70. https://doi.org/10.2217/pme.10.75.

17. Fang H, Hui Q, Lynch J, et al. Harmonizing genetic ancestry and self-identified race/ethnicity in genome-wide association studies. *Am J Hum Genet.* 2019;105(4):763−772. https://doi.org/10.1016/j.ajhg.2019.08.012.

18. Ross PT, Hart-Johnson T, Santen SA, Zaidi NLB. Considerations for using race and ethnicity as quantitative variables in medical education research. *Perspect Med Educat.* 2020;9(5):318−323. https://doi.org/10.1007/s40037-020-00602-3.

19. Mersha TB, Abebe T. Self-reported race/ethnicity in the age of genomic research: its potential impact on understanding health disparities. *Hum Genom.* 2015;9(1):1. https://doi.org/10.1186/s40246-014-0023-x.

20. Cavalli-Sforza LL, Feldman MW. The application of molecular genetic approaches to the study of human evolution. *Nat Genet.* 2003;33:266−275.

21. Kotze MJ, Langenhoven E, Theart L, Loubser O, Micklem A, Oosthuizen CJ. Recurrent LDL-receptor mutation causes familial hypercholesterolaemia in South African coloureds and Afrikaners. *South African Medical Journal.* 1995;85(5):357−361.

22. Yang G, Alarcon C, Friedman P, et al. The role of global and local ancestry on clopidogrel response in African Americans. In: *Pacific Symposium on Biocomputing Pacific Symposium on Biocomputing*. 2023: 221−232, 28.

23. Uren C, Kim M, Martin AR, et al. Fine-scale human population structure in southern Africa reflects eco-geographic boundaries. *Genetics*. 2016;204(1):303−314. https://doi.org/10.1534/genetics.116.187369.

24. Uren C, Möller M, van Helden PD, Henn BM, Hoal EG. Population structure and infectious disease risk in southern Africa. *Mol Genet Genom : MGG*. 2017;292(3):499−509. https://doi.org/10.1007/s00438-017-1296-2.

25. Loubser O, Marais AD, Kotze MJ, et al. Founder mutations in the LDL receptor gene contribute significantly to the familial hypercholesterolemia phenotype in the indigenous South African population of mixed ancestry. *Clin Genet*. 1999;55(5):340−345.

26. Luo Y, Suliman S, Asgari S, et al. Early progression to active tuberculosis is a highly heritable trait driven by 3q23 in Peruvians. *Nat Commun*. 2019;10(1):3765. https://doi.org/10.1038/s41467-019-11664-1.

27. Maples BK, Gravel S, Kenny EE, Bustamante CD. RFMix: a discriminative modeling approach for rapid and robust local-ancestry inference. *Am J Hum Genet*. 2013;93(2):278−288. https://doi.org/10.1016/j.ajhg.2013.06.020.

28. Uren C, Hoal EG, Möller M. Putting RFMix and ADMIXTURE to the test in a complex admixed population. *BMC Genet*. 2020;21(1):40. https://doi.org/10.1186/s12863-020-00845-3.

29. Winkler CA, Nelson GW, Smith MW. Admixture mapping comes of age. *Annu Rev Genom Hum Genet*. 2010;11:65−89. https://doi.org/10.1146/annurev-genom-082509-141523.

30. Gordon DK, Hussain M, Kumar P, Khan S, Khan S. The sickle effect: the silent titan affecting glycated hemoglobin reliability. *Cureus*. 2020;12(8):e9685. https://doi.org/10.7759/cureus.9685.

31. Zemlin AE, Matsha TE, Hassan MS, Erasmus RT. HbA1c of 6.5% to diagnose diabetes mellitus – does it work for us? – the Bellville South Africa study. *PLoS One*. 2011;6(8):e22558. https://doi.org/10.1371/journal.pone.0022558.

32. Kengne AP, Matsha TE, Sacks DB, Zemlin AE, Erasmus RT, Sumner AE. And glycated albumin improves detection of dysglycaemia in mixed-ancestry South Africans. *EClin Med*. 2022;48:101443. https://doi.org/10.1016/j.eclinm.2022.101443.

33. Hivert M-F, Christophi CA, Jablonski KA, et al. Genetic ancestry markers and difference in A1c between African American and white in the diabetes prevention program. *J Clin Endocrinol Metab*. 2019;104(2): 328−336. https://doi.org/10.1210/jc.2018-01416.

34. Skinner S, Diaw M, Ndour MM, et al. Evaluation of agreement between hemoglobin A1c, fasting glucose, and fructosamine in senegalese individuals with and without sickle-cell trait. *PLoS One*. 2019;14(2). https://doi.org/10.1371/journal.pone.0212552. e0212552.

35. Ozarda Y, Sikaris K, Streichert T, Macri J. Distinguishing reference intervals and clinical decision limits - a review by the IFCC Committee on reference intervals and decision limits. *Crit Rev Clin Lab Sci*. 2018; 55(6):420−431. https://doi.org/10.1080/10408363.2018.1482256.

36. Arneson WL, Arneson DL. Current methods for routine clinical laboratory testing of vitamin D levels. *Lab Med*. 2013;44(1):e38−e42. https://doi.org/10.1309/LMONQZQ27TIN7XFS.

37. Dhaliwal R, Islam S, Mikhail M, Ragolia L, Aloia JF. Effect of vitamin D on bone strength in older African Americans: a randomized controlled trial. *Osteoporos Int: J Estab Res Cooperat Europ Foundat Osteopor National Osteopor Foundat U S A*. 2020;31(6):1105−1114. https://doi.org/10.1007/s00198-019-05275-1.

38. Thamattoor A. Race/ethnicity differences in vitamin D levels and impact on cardiovascular disease, bone health, and oral health. *medRxiv*. 2021. https://doi.org/10.1101/2021.01.02.21249149.

39. ODS. Vitamin D Fact Sheet for Health Professionals. https://ods.od.nih.gov/factsheets/VitaminD-HealthProfessional/. Accessed May 31, 2023.

40. Okunola AO, Baatjes KJ, Zemlin AE, et al. Pathology-supported genetic testing for the application of breast cancer pharmacodiagnostics: family counselling, lifestyle adjustments and change of medication. *Expert Rev Mol Diagn.* 2023;23(5):431−443. https://doi.org/10.1080/14737159.2023.2203815.

41. Baatjes K, Peeters A, McCaul M, et al. CYP19A1 rs10046 pharmacogenetics in postmenopausal breast cancer patients treated with aromatase inhibitors: one-year follow-up. *Curr Pharmaceut Des.* 2020;26(46): 6007−6012. https://doi.org/10.2174/1381612826666200908141858.

42. Browning SR, Grinde K, Plantinga A, et al. Local ancestry inference in a large US-based hispanic/latino study: hispanic community health study/study of latinos (HCHS/SOL). *G3 (Bethesda, Md).* 2016;6(6): 1525−1534. https://doi.org/10.1534/g3.116.028779.

43. Damier P, Hirsch EC, Agid Y, Graybiel AM. The substantia nigra of the human brain. II. Patterns of loss of dopamine-containing neurons in Parkinson's disease. *Brain: J Neurol.* 1999;122(Pt 8):1437−1448. https://doi.org/10.1093/brain/122.8.1437.

44. Jankovic J. Parkinson's disease: clinical features and diagnosis. *J Neurol Neurosurg Psychiatr.* 2008;79(4): 368−376. https://doi.org/10.1136/jnnp.2007.131045.

45. Chaudhuri KR, Schapira AHV. Non-motor symptoms of Parkinson's disease: dopaminergic pathophysiology and treatment. *Lancet Neurol.* 2009;8(5):464−474. https://doi.org/10.1016/S1474-4422(09)70068-7.

46. Poewe W. Non-motor symptoms in Parkinson's disease. *Eur J Neurol.* 2008;15(Suppl 1):14−20. https://doi.org/10.1111/j.1468-1331.2008.02056.x.

47. Gibb WR, Lees AJ. The relevance of the Lewy body to the pathogenesis of idiopathic Parkinson's disease. *J Neurol Neurosurg Psychiatr.* 1988;51(6):745−752. https://doi.org/10.1136/jnnp.51.6.745.

48. Rizzo G, Copetti M, Arcuti S, Martino D, Fontana A, Logroscino G. Accuracy of clinical diagnosis of Parkinson disease: a systematic review and meta-analysis. *Neurology.* 2016;86(6):566−576. https://doi.org/10.1212/WNL.0000000000002350.

49. Pagano G, Niccolini F, Politis M. Imaging in Parkinson's disease. *Clin Med.* 2016;16(4):371−375. https://doi.org/10.7861/clinmedicine.16-4-371.

50. Hamid E, Ayele BA, Massi DG, et al. Availability of therapies and services for Parkinson's disease in africa: a continent-wide survey. *Mov Disord.* 2021;36(10):2393−2407. https://doi.org/10.1002/mds.28669.

51. Vaidyanathan R, Soon RH, Zhang P, Jiang K, Lim CT. Cancer diagnosis: from tumor to liquid biopsy and beyond. *Lab Chip.* 2018;19(1):11−34. https://doi.org/10.1039/C8LC00684A.

52. Parnetti L, Gaetani L, Eusebi P, et al. CSF and blood biomarkers for Parkinson's disease. *Lancet Neurol.* 2019;18(6):573−586. https://doi.org/10.1016/S1474-4422(19)30024-9.

53. Anney RJL, Ripke S, Anttila V, et al. Meta-analysis of GWAS of over 16,000 individuals with autism spectrum disorder highlights a novel locus at 10q24.32 and a significant overlap with schizophrenia. *Mol Autism.* 2017;8(1):21. https://doi.org/10.1186/s13229-017-0137-9.

54. Blauwendraat C, Faghri F, Pihlstrom L, et al. NeuroChip, an updated version of the NeuroX genotyping platform to rapidly screen for variants associated with neurological diseases. *Neurobiol Ag.* 2017;57. https://doi.org/10.1016/j.neurobiolaging.2017.05.009, 247.e9-247.e13.

55. Farhan SMK, Dilliott AA, Ghani M, et al. The ONDRISeq panel: custom-designed next-generation sequencing of genes related to neurodegeneration. *NPJ Genom Med.* 2016;1:16032. https://doi.org/10.1038/npjgenmed.2016.32.

56. Yu AC-S, Yim AK-Y, Chan AY-Y, et al. A targeted gene panel that covers coding, non-coding and short tandem repeat regions improves the diagnosis of patients with neurodegenerative diseases. *Front Neurosci.* 2019;13:1324. https://doi.org/10.3389/fnins.2019.01324.

57. Müller-Nedebock AC, Komolafe MA, Fawale MB, et al. Copy number variation in Parkinson's disease: an update from sub-saharan africa. *Mov Disord: Offic J Movem Dis Soci.* 2021;36(10):2442−2444. https://doi.org/10.1002/mds.28710.

58. Sebate B, Cuttler K, Cloete R, et al. Prioritization of candidate genes for a South African family with Parkinson's disease using in-silico tools. *PLoS One*. 2021;16(3). https://doi.org/10.1371/journal.pone.0249324. e0249324.

59. Farrow SL, Cooper AA, O'Sullivan JM. Redefining the hypotheses driving Parkinson's diseases research. *NPJ Parkinson's Dis*. 2022;8(1):45. https://doi.org/10.1038/s41531-022-00307-w.

60. Mestre TA, Fereshtehnejad S-M, Berg D, et al. Parkinson's disease subtypes: critical appraisal and recommendations. *J Parkinsons Dis*. 2021;11(2):395−404. https://doi.org/10.3233/JPD-202472.

61. Anwarullah AM, Badshah M, et al. Further evidence for the association of CYP2D6*4 gene polymorphism with Parkinson's disease: a case control study. *Gene Environ: Offic J Jap Environ Mutag Soci*. 2017;39:18. https://doi.org/10.1186/s41021-017-0078-8.

62. van der Merwe N, Bouwens CSH, Pienaar R, et al. CYP2D6 genotyping and use of antidepressants in breast cancer patients: test development for clinical application. *Metab Brain Dis*. 2012;27(3):319−326. https://doi.org/10.1007/s11011-012-9312-z.

63. Sandin S, Lichtenstein P, Kuja-Halkola R, Larsson H, Hultman CM, Reichenberg A. The familial risk of autism. *JAMA*. 2014;311(17):1770−1777. https://doi.org/10.1001/jama.2014.4144.

64. Waterhouse L, Morris R, Allen D, et al. Diagnosis and classification in autism. *J Autism Dev Disord*. 1996;26(1):59−86. https://doi.org/10.1007/BF02276235.

65. Leibu E, First MB. Dsm-5. In: *Mount Sinai Expert Guides*. John Wiley and Sons, Ltd; 2016:1−8. https://onlinelibrary.wiley.com/doi/abs/10.1002/9781118654231.ch1. Accessed November 30, 2022.

66. Fetit R, Hillary RF, Price DJ, Lawrie SM. The neuropathology of autism: a systematic review of post-mortem studies of autism and related disorders. *Neurosci Biobehav Rev*. 2021;129:35−62. https://doi.org/10.1016/j.neubiorev.2021.07.014.

67. Masi A, DeMayo MM, Glozier N, Guastella AJ. An overview of autism spectrum disorder, heterogeneity and treatment options. *Neurosci Bull*. 2017;33(2):183−193. https://doi.org/10.1007/s12264-017-0100-y.

68. Grove J, Ripke S, Als TD, et al. Identification of common genetic risk variants for autism spectrum disorder. *Nat Genet*. 2019;51(3):431−444. https://doi.org/10.1038/s41588-019-0344-8.

69. Liu X, Shimada T, Otowa T, et al. Genome-wide association study of autism spectrum disorder in the east Asian populations. *Autism Res*. 2016;9(3):340−349. https://doi.org/10.1002/aur.1536.

70. Satterstrom FK, Kosmicki JA, Wang J, et al. Large-scale exome sequencing study implicates both developmental and functional changes in the neurobiology of autism. *Cell*. 2020;180(3):568. https://doi.org/10.1016/j.cell.2019.12.036, 584.e23.

71. Gadow KD, DeVincent C, Schneider J. Predictors of psychiatric symptoms in children with an autism spectrum disorder. *J Autism Dev Disord*. 2008;38(9):1710−1720. https://doi.org/10.1007/s10803-008-0556-8.

72. Sikora DM, Vora P, Coury DL, Rosenberg D. Attention-deficit/hyperactivity disorder symptoms, adaptive functioning, and quality of life in children with autism spectrum disorder. *Pediatrics*. 2012;130(2):S91−S97. https://doi.org/10.1542/peds.2012-0900G.

73. Ruparelia K, Abubakar A, Badoe E, et al. Autism spectrum disorders in africa: current challenges in identification, assessment, and treatment: a report on the international child neurology association meeting on ASD in africa, Ghana, April 3-5, 2014. *J Child Neurol*. 2016;31(8):1018−1026. https://doi.org/10.1177/0883073816635748.

74. Franz L, Chambers N, von Isenburg M, de Vries PJ. Autism spectrum disorder in sub-saharan africa: a comprehensive scoping review. *Autism Res*. 2017;10(5):723−749. https://doi.org/10.1002/aur.1766.

75. Miller DT, Adam MP, Aradhya S, et al. Consensus statement: chromosomal microarray is a first-tier clinical diagnostic test for individuals with developmental disabilities or congenital anomalies. *Am J Hum Genet*. 2010;86(5):749−764. https://doi.org/10.1016/j.ajhg.2010.04.006.

76. Accelerating Excellence in Science in Africa AESA. A framework for the implementation of genomic medicine for public. *AAS Open Res.* 2021;4(9). https://doi.org/10.21955/aasopenres.1115149.

77. Jongeneel CV, Kotze MJ, Bhaw-Luximon A, et al. A view on genomic medicine activities in africa: implications for policy. *Front Genet.* 2022;13:769919. https://doi.org/10.3389/fgene.2022.769919.

78. Weigelt B, Geyer FC, Reis-Filho JS. Histological types of breast cancer: how special are they? *Mol Oncol.* 2010;4(3):192−208. https://doi.org/10.1016/j.molonc.2010.04.004.

79. *Worldwide Cancer Data | World Cancer Research Fund International.* WCRF International; 2020. https://www.wcrf.org/cancer-trends/worldwide-cancer-data/. Accessed November 15, 2022.

80. De Jager LJ, Schubert PT, Baatjes K, Conradie W, Edge J. A comparison of invasive lobular carcinoma with other invasive breast cancers at Tygerberg Academic Hospital. *South African J Surg Suid-Afrikaanse Tydskrif Vir Chirurgie.* 2022;60(3):176−181. https://doi.org/10.17159/2078-5151/SAJS3768.

81. Perou CM, Sørlie T, Eisen MB, et al. Molecular portraits of human breast tumours. *Nature.* 2000;406(6797): 747−752. https://doi.org/10.1038/35021093.

82. Cardoso F, van't Veer LJ, Bogaerts J, et al. 70-Gene signature as an aid to treatment decisions in early-stage breast cancer. *N Engl J Med.* 2016;375(8):717−729. https://doi.org/10.1056/NEJMoa1602253.

83. Grant KA, Apffelstaedt JP, Wright CA, et al. MammaPrint Pre-screen Algorithm (MPA) reduces chemotherapy in patients with early-stage breast cancer. *South African Med J.* 2013;103(8):522−526. https://doi.org/10.7196/samj.7223.

84. Harbeck N, Gnant M. Breast cancer. *Lancet.* 2017;389(10074):1134−1150. https://doi.org/10.1016/S0140-6736(16)31891-8.

85. Myburgh EJ, de Jager JJ, Murray E, Grant KA, Kotze MJ, de Klerk H. The cost impact of unselective vs selective MammaPrint testing in early-stage breast cancer in southern Africa. *Breast.* 2021;59:87−93. https://doi.org/10.1016/j.breast.2021.05.010.

86. Mampunye I, Grant KA, Peeters AV, et al. MammaPrint risk score distribution in South African breast cancer patients with the pathogenic BRCA2 c.7934delG founder variant: towards application of genomic medicine at the point-of-care. *Breast.* 2021;56:S31. https://doi.org/10.1016/S0960-9776(21)00120-X.

87. Mampunye L, van der Merwe NC, Grant KA, et al. Pioneering BRCA1/2 point-of-care testing for integration of germline and tumor genetics in breast cancer risk management: a vision for the future of translational pharmacogenomics. *Front Oncol.* 2021;11:619817. https://doi.org/10.3389/fonc.2021.619817.

88. Lammert J, Grill S, Kiechle M. Modifiable lifestyle factors: opportunities for (hereditary) breast cancer prevention - a narrative review. *Breast Care.* 2018;13(2):109−114. https://doi.org/10.1159/000488995.

89. Van der Merwe N. *Development and Application of a Pathology Supported Pharmacogenetic Test for Improved Clinical Management of South African Patients with Breast Cancer and Associated Comorbidities*; 2016. https://scholar.sun.ac.za:443/handle/10019.1/98481. Accessed November 16, 2022.

90. Esserman LJ, Yau C, Thompson CK, et al. Use of molecular tools to identify patients with indolent breast cancers with ultralow risk over 2 decades. *JAMA Oncol.* 2017;3(11):1503−1510. https://doi.org/10.1001/jamaoncol.2017.1261.

91. Kotze M, Malan J, Pienaar R, Apffelstaedt J. The role of molecular genetic testing in modern breast heath management. *S Afr Fam Pract.* 2005;47(9):38−40. https://doi.org/10.1080/20786204.2005.10873286.

92. van der Merwe NC, Ntaita KS, Stofberg H, Combrink HM, Oosthuizen J, Kotze MJ. Implementation of multigene panel testing for breast and ovarian cancer in South Africa: a step towards excellence in oncology for the public sector. *Front Oncol.* 2022;12:938561. https://doi.org/10.3389/fonc.2022.938561.

93. Cardoso F, Curigliano G. The PROMISe to increase precision in adjuvant therapy for early breast cancer: to "Type" or to "Print". *NPJ Breast Canc.* 2018;4:12. https://doi.org/10.1038/s41523-018-0064-8.

94. Hoskins KF, Danciu OC, Ko NY, Calip GS. Association of race/ethnicity and the 21-gene recurrence score with breast cancer-specific mortality among US women. *JAMA Oncol.* 2021;7(3):370−378. https://doi.org/10.1001/jamaoncol.2020.7320.

95. Essien E-O, Rali P, Mathai SC. Pulmonary embolism. *Med Clin*. 2019;103(3):549−564. https://doi.org/10.1016/j.mcna.2018.12.013.

96. Konecny F. Inherited trombophilic states and pulmonary embolism. *J Res Med Sci: Offic J Isfahan University Med Sci*. 2009;14(1):43−56.

97. Nizankowska-Mogilnicka E, Adamek L, Grzanka P, et al. Genetic polymorphisms associated with acute pulmonary embolism and deep venous thrombosis. *Eur Respir J*. 2003;21(1):25−30.

98. Zöller B, Svensson PJ, Dahlbäck B, Lind-Hallden C, Hallden C, Elf J. Genetic risk factors for venous thromboembolism. *Expet Rev Hematol*. 2020;13(9):971−981. https://doi.org/10.1080/17474086.2020.1804354.

99. Doherty S. Pulmonary embolism an update. *Aust Fam Physician*. 2017;46(11):816−820.

100. Konstantinides SV, Meyer G, Becattini C, et al. ESC Guidelines for the diagnosis and management of acute pulmonary embolism developed in collaboration with the European Respiratory Society (ERS). *Eur Heart J*. 2019;41(4):543−603. https://doi.org/10.1093/eurheartj/ehz405.

101. Kheiri B, Abdalla A, Haykal T, et al. Meta-analysis of genotype-guided versus standard dosing of vitamin K antagonists. *Am J Cardiol*. 2018;121(7):879−887. https://doi.org/10.1016/j.amjcard.2017.12.023.

102. Mostbauer H, Nishkumay O, Rokyta O, Vavryniuk V. Warfarin resistance: possibilities to solve this problem. A case report. *J Int Med Res*. 2022;50(6). https://doi.org/10.1177/03000605221103959, 3000605221103959.

103. Kotze MJ, Lückhoff HK, Peeters AV, et al. Genomic medicine and risk prediction across the disease spectrum. *Crit Rev Clin Lab Sci*. 2015;52(3):120−137. https://doi.org/10.3109/10408363.2014.997930.

104. Talevich E, Shain AH, Botton T, Bastian BC. CNVkit: genome-wide copy number detection and visualization from targeted DNA sequencing. *PLoS Comput Biol*. 2016;12(4). https://doi.org/10.1371/journal.pcbi.1004873. e1004873.

105. Lückhoff HK, Kidd M, van Rensburg SJ, van Velden DP, Kotze MJ. Apolipoprotein E genotyping and questionnaire-based assessment of lifestyle risk factors in dyslipidemic patients with a family history of Alzheimer's disease: test development for clinical application. *Metab Brain Dis*. 2016;31(1):213−224.

106. Delport D, Schoeman R, van der Merwe N, et al. Significance of dietary folate intake, homocysteine levels and MTHFR 677 C>T genotyping in South African patients diagnosed with depression: test development for clinical application. *Metab Brain Dis*. 2014;29(2):377−384. https://doi.org/10.1007/s11011-014-9506-7.

107. Kotze MJ, van Velden DP, Botha K, et al. Pathology-supported genetic testing directed at shared disease pathways for optimized health in later life. *Pers Med*. 2013;10(5):497−507. https://doi.org/10.2217/pme.13.43.

108. García-Estévez L, Cortés J, Pérez S, Calvo I, Gallegos I, Moreno-Bueno G. Obesity and breast cancer: a paradoxical and controversial relationship influenced by menopausal status. *Front Oncol*. 2021;11:705911. https://doi.org/10.3389/fonc.2021.705911.

109. Renehan AG, Tyson M, Egger M, Heller RF, Zwahlen M. Body-mass index and incidence of cancer: a systematic review and meta-analysis of prospective observational studies. *Lancet*. 2008;371(9612):569−578. https://doi.org/10.1016/S0140-6736(08)60269-X.

110. Bardia A, Arieas ET, Zhang Z, et al. Comparison of breast cancer recurrence risk and cardiovascular disease incidence risk among postmenopausal women with breast cancer. *Breast Cancer Res Treat*. 2012;131(3):907−914. https://doi.org/10.1007/s10549-011-1843-1.

Implementing population pharmacogenomics: Tailoring drug therapy for diverse populations

Kariofyllis Karamperis[1,2,3] **and George P. Patrinos**[1,4,5,6]

[1]*Laboratory of Pharmacogenomics and Individualized Therapy, Department of Pharmacy, School of Health Sciences, University of Patras, Patras, Greece;* [2]*Group of Algorithms for Population Genomics, Department of Genetics, Institut de Biologia Evolutiva, IBE, (CSIC-Universitat Pompeu Fabra), Barcelona, Spain;* [3]*The Golden Helix Foundation, London, UK;* [4]*Department of Genetics and Genomics, College of Medicine and Health Sciences, United Arab Emirates University, Al-Ain, Abu Dhabi, United Arab Emirates;* [5]*Zayed Center for Health Sciences, United Arab Emirates University, Al-Ain, Abu Dhabi, United Arab Emirates;* [6]*Clinical Bioinformatics Unit, Department of Pathology, Faculty of Medicine and Health Sciences, Erasmus University Medical Center, Rotterdam, The Netherlands*

Personalized medicine: "One size does not fit all"

Recent advancements in genomics, biotechnology, and data analytics have significantly propelled the evolution of personalized medicine, an emerging approach that exploits an individual's genetic profile for tailoring disease prevention and treatment.[1] Although the term "personalized medicine," also known as precision medicine, first appeared in 1999 in a Wall Street Journal article titled "New Era of Personalized Medicine: Targeting Drugs for Each Unique Genetic Profile," the concept has roots that trace back thousands of years to Hippocrates in the 4th century B.C. He famously noted, *"It is far more important to know what person the disease has than what disease the person has."*[2,3] This quote underscores the timeless pursuit of understanding human physiology and inter-individual variability resulting in the optimization of modern medicine. Over the years, current medicine has followed an empirical approach relying on observation, experimentation, and evidence-based practices for diagnosis, treatment, and prevention of a disease or other related conditions. In other words, standardized approaches are not personalized to each patient follow a "one-size-fits-all" approach. Notably, Giuseppe A. Giacomini, a prominent figure in pharmacogenetics and pharmacogenomics, recently emphasized the limitations of this approach, stating, *"It is unthinkable that selecting drugs for individual patients remains an empirical exercise."*[4] This perspective is widely shared among the scientific community, underscoring the need for exploring new methodologies or enhancing existing ones.

On the contrary, personalized medicine marks a significant shift from the traditional one-size-fits-all model to a more precise approach that considers individual variability in genes, response to medications, and disease susceptibility.[5,6] It recognizes that an individual's unique biological, genetic, and environmental factors can influence their response to treatment.[7,8] By leveraging advancements, such as genomic sequencing, biomarker analysis, and sophisticated diagnostic tools, healthcare professionals can tailor therapies to align with the specific characteristics of each patient to enable more precise, effective, and individualized care.[9,10] Nowadays, this revolutionary approach holds immense promise for a wide range of applications across various fields of healthcare, including cancer treatment, cardiovascular diseases, psychiatric disorders, rare genetic conditions, diabetes management, and more. Apart from these, it has also shown significant advancements in pharmacogenomics by personalizing drug treatment to deliver *"the right drug, to the right patient, at the right time."*[10] This chapter will delve into this groundbreaking field and, to a great extent, the emerging concept of population pharmacogenomics.

Pharmacogenomics: The right drug, to the right patent, at the right time

Pharmacogenomics (PGx), a key component of personalized medicine, combines genomics and pharmacology to study how genetic variations influence an individual's response to drugs. By analyzing genetic variations in genes that affect drug metabolism, efficacy, and safety, pharmacogenomics aims to optimize drug therapy, enhancing effectiveness and reducing the risk of adverse effects.[11,12] The majority of these genetic variations are primarily located in genes encoding drug-metabolizing enzymes (i.e., CYP450), drug transporters (such as ATP-binding cassette subfamilies, solute carrier transporters[SLC]), drug receptors (G-protein coupled receptors [GPCRs], ion channel receptors), or other drug targets (angiotensin-converting enzyme [ACE]).[11,13−16] Depending on the amino acid substitution and its effect on protein structure, these genetic variations can significantly impact drug efficacy, potentially leading to reduced therapeutic outcomes or even life-threatening conditions for the patient. It is important to note that in the field of pharmacogenomics, the relevant genes are often referred to as pharmacogenes or ADME genes, where ADME stands for the key stages of pharmacokinetics: absorption, distribution, metabolism, and excretion.[17] Nowadays, a significant portion of research focuses on pharmacogenes to elucidate why certain individuals exhibit a lack of therapeutic response, require dose adjustments for specific medications, or experience adverse drug reactions.[18] For instance, the CYP1A2, CYP2D6, CYP2C19, and CYP3A4 enzymes, belonging to the CYP1, CYP2, and CYP3 families, respectively, are extensively involved in the biotransformation (metabolism) of many foreign compounds, accounting for approximately 70%−80% of all drugs used in clinical practice.[19,20] Additionally, most genes encoding these enzymes tend to be highly polymorphic among individuals, which can result in variations in each patient's metabolizer status.[21] In some cases, these genetic differences can cause certain individuals to metabolize drugs too slowly or too quickly, increasing the risk of drug toxicity or reduced therapeutic efficacy. Consequently, dosage adjustments or the selection of alternative medications may be necessary to ensure safe and effective treatment.[22] Therefore, sequencing analysis is crucial for exploring and uncovering a plethora of genetic regions involved in pharmacokinetics and pharmacodynamics, and for deciphering their impact on commonly prescribed drugs.[10,23−25]

Adverse drug reactions (ADRs) represent a global public health challenge, significantly affecting both patients and the healthcare systems, with a high incidence that is expected to increase in the near future. According to the World Health Organization (WHO)[26] and Centers for Disease Control and Prevention (CDC),[27] ADRs account for over a million emergency department visits in the USA annually and rank as the sixth leading cause of death in the country. In fact, approximately 350,000 patients require hospitalization for further treatment following emergency visits due to ADRs.[28] Individualized therapy and the implementation of pharmacogenomics are deemed essential, as integrating personalized medicine into drug treatment not only improves patient care by tailoring interventions but also contributes to broader objectives such as enhancing treatment outcomes, reducing healthcare costs, and facilitating the development of novel therapeutics through more targeted clinical trials.[29,30] Given this context, pharmacogenomics holds immense promise for reshaping personalized medicine by minimizing these drug-related events and tailoring drug treatments simultaneously. However, the clinical implementation of pharmacogenomics remains challenging due to significant limitations that hinder its full integration into healthcare systems across the globe.[31]

The present landscape of pharmacogenomics: Necessity or luxury?

Recently, there has been a strong debate around pharmacogenomics, with discussions focusing on whether its integration into clinical practice is an essential advancement for personalized medicine or a luxury that may not be feasible for widespread implementation in the near future. There is substantial clinical evidence supporting the notion that genetic variation can significantly influence a patient's response to medications, with recommendations continuously being updated by regulatory bodies. Characteristically, the U.S. Food and Drug Administration (FDA)[32] has approved over 300 drug labels containing pharmacogenetic recommendations, while the European Medicines Agency (EMA)[33] has approved more than 150 such labels.[31] Consequently, prior to drug treatment, pharmacogenomics holds the potential to offer numerous benefits, acting as a predictive tool and enabling genome-guided treatments worldwide.[34–36] The concept has garnered considerable attention in fields such as cardiology, psychiatry, and oncology, and is increasingly recognized for its potential preemptive applications in areas such as transplantation, pain management, and beyond. Notably, recent findings from the first prospective European pharmacogenomics clinical study were highly encouraging, revealing a 30% reduction in side effects compared to those patients who did not receive genome-guided treatment.[37]

Current pharmacogenomics evidence is accessible through publicly available databases such as Pharmacogenomics Knowledge Base (PharmGKB),[38] ClinVar,[39] Clinical Pharmacogenetics Implementation Consortium (CPIC),[40] and PharmVar,[41] which serve as repositories for a wide range of genetic variations related to drug response and other relevant phenotypes. More specifically, these databases compile detailed information on how particular genetic profiles affect an individual's response to medications, including aspects such as metabolism rates, efficacy, and potential adverse effects. By aggregating and organizing this extensive genetic data, these resources support research, enhance clinical decision-making, and promote the adoption of personalized medicine. For example, PharmGKB provides thousands of clinical annotations and pharmacogenomic guidelines, including those published by regulatory agencies such as the CPIC[42] and the Dutch Pharmacogenetics Working Group (DPWG).[43] These annotations associate single nucleotide variants/polymorphisms (SNVs/SNPs) with star alleles for various drugs and drug-related phenotypes. Star alleles are described as the phenotypic outcome of a genetic variant in terms of functionality and might correspond to a combination of genetic variants.[44]

Despite its significant promise for personalized medicine, pharmacogenomics still faces limitations that must be addressed before its clinical implementation and integration into healthcare systems.[31] The implementation of next-generation sequencing (NGS) technologies in pharmacogenomics encounters significant challenges, primarily due to the high costs associated with both the sequencing technology and the subsequent data analysis. NGS platforms and their application remain prohibitively expensive for routine clinical use.[45] Additionally, the requirement for substantial computational infrastructure and highly skilled personnel, including experienced physicians and healthcare personnel , to analyze, process, and interpret extensive volumes of sequencing data further drives up the cost.[46] Therefore, the incorporation of pharmacogenetic testing, the interpretation of results, physician education, and patient acceptance must also be considered. As a result, integrating pharmacogenomic information into clinical practice might not be accessible in some regions, including developing countries due to the limited healthcare infrastructure and data management systems.[47,48] Furthermore, many drugs can lead to complex interactions influenced by a variety of genetic variants, environmental factors, and lifestyle variables, as previously mentioned. Understanding these intricate relationships between genes and drug responses, known as drug-gene interactions, requires extensive research and thorough data analysis. While some drug-gene interactions are well-established, comprehensive evidence is still lacking for many others. The insufficiency of data and the presence of conflicting findings in pharmacogenomic research limit the development of robust guidelines for clinical decision-making.[49,50]

Among other challenges, a significant constraint involves individual characteristics, such as age, gender, and, most importantly, ethnicity, which are often not adequately considered. For instance, it has been extensively reported that, in addition to inter-individual variability, population differences among individuals have been also observed, hindering the recommended drug and dosage predictions across different populations.[51,52] Exploring and identifying population-level variations prior to the implementation of pharmacogenomics can reveal distinct geographical patterns that may be utilized for better stratification of individuals.[53] This emerging concept, known as population pharmacogenomics, will be further discussed in the following paragraphs.

Introducing population pharmacogenomics

In recent years, it has been extensively reported that genetic variation arises from a complex interplay of genetic, environmental, and epigenetic factors, influencing traits, disease susceptibility, drug responses, and various physiological processes. Interestingly, numerous studies have underscored that apart from the inter-individual variability, in most cases, inter- or intra-population variability is also prevalent.[24,52–55] More specifically, certain populations display greater genetic heterogeneity compared to others, indicating the potential to discern underlying genetic patterns. Based on the latest findings, on average, 85% of genetic variation exists within local populations, 7% between local populations on the same continent, whereas 8% exists between large groups living on different continents. Nevertheless, because of the shared ancestry of all humans, only a limited number of genetic variants exhibit significant differences between populations. However, identifying rare variants (MAF = <1%) or novel variants for a particular population allows us to predict a pattern related to a disease or a phenotype (Fig. 8.1).[56–58] It is well known that in population genetics, the goal is to understand how genetic traits and variations are distributed within and between populations, how these

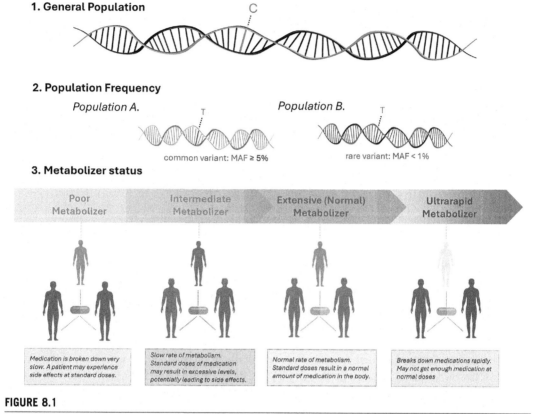

1. General Population

C

2. Population Frequency

Population A.

T

common variant: MAF ≥ 5%

Population B.

T

rare variant: MAF < 1%

3. Metabolizer status

| Poor Metabolizer | Intermediate Metabolizer | Extensive (Normal) Metabolizer | Ultrarapid Metabolizer |

Medication is broken down very slow. A patient may experience side effects at standard doses.

Slow rate of metabolism. Standard doses of medication may result in excessive levels, potentially leading to side effects.

Normal rate of metabolism. Standard doses result in a normal amount of medication in the body.

Breaks down medications rapidly. May not get enough medication at normal doses

FIGURE 8.1

Schematic representation highlighting the importance of integrating population pharmacogenomics into personalized medicine.

distributions change over time through processes such as mutation, natural selection, genetic drift, and gene flow, and how these genetic changes contribute to the evolution of species.[59,60] Similarly, population pharmacogenomics is specifically focused on studying the relationship between genetic variations within populations and their responses to drugs within different populations. It merges concepts from pharmacogenetics, which concentrates on how individual genetic differences affect drug responses, with population genetics, which studies genetic diversity within and between different populations[61,62] to understand how these genetic differences impact drug metabolism, efficacy, and potential side effects at a population level. By analyzing large cohorts from diverse ethnic backgrounds, researchers seek to identify pharmacogenetic and pharmacogenomic markers linked to variability in drug responses. These markers may include SNPs, copy number variations (CNVs), star alleles, and other genetic variants that influence drug metabolism, transport, or other drug-target interactions.[47,61] Compared to the standardized individual-based approach in pharmacogenomics, a population-centered approach can help predict similarities and differences between populations, potentially altering the risk-benefit profiles in specific groups.[63] In simple terms, by understanding the genetic background of a group of individuals between populations, we can detect the minor allele frequency and then stratify populations by which of them tend to have a relatively higher risk or lower response or susceptibility to ADRs to a certain drug.[64,65] Please note that the term *"population"* refers to a group of interbreeding individuals who share a common set of genes within a defined geographic area and are often characterized by their genetic composition and variation. Geographical features such as culture, language, and tradition do not always reflect genetic background and may lead to misleading speculations.[66] To date, large-scale pharmacogenomic studies have emphasized the importance of considering individual characteristics, especially in pharmacogenomics, given the range of benefits that will be discussed in the following sections.

Pharmacogenomics across diverse populations

Examining the prevalence of allele frequencies across diverse populations reveals significant variations in pharmacogenetic markers that can impact drug efficacy and safety. Independent studies from Latin America, Southeast Asia, and Africa have identified hundreds of clinically actionable pharmacogenomic variants with statistically significant allele frequencies that affect drug treatment, revealing notable differences compared to other global populations. In particular, Latin America, a highly diverse region encompassing a wide range of genetic backgrounds, these results can contribute to delineating significant variations in pharmacogenomic markers across different populations within the region. Similar findings have been reported in European populations.[52,67–70] More specifically, observations from 18 European populations have revealed significant inter-population differences in pharmacogenomic biomarker allele frequencies, particularly in seven clinically actionable biomarkers. Notable differences were identified in the *CYP2C9, CYP2C19, CYP3A5, VKORC1, SLCO1B1,* and *TPMT* pharmacogenes, underscoring the potential importance of considering these variations in clinical practice. These studies also highlight the significance of drug-gene interactions in the context of cardiovascular drugs, antipsychotics, antidepressants, and antineoplastic agents. Certain populations exhibit higher allele frequencies for specific drug categories while showing lower frequencies for others, emphasizing the importance for tailored treatments based on population-specific genetic profiles.[50,71,72] The observed diversity in these regions, characterized by high levels of genetic admixture, supports the presence of significant population differences. Notably, these studies have uncovered inter-population variations, suggesting that this approach could facilitate the development of population-specific pharmacogenomic panels and guidelines, crucial for accurately predicting drug selection and response or dosage adjustment, when necessary.[49,73] Moving forward, inter-population variability has also been observed among Qatari subpopulations in the Middle East and Indians in South Asia, enhancing the critical need for a population-specific approach.[68,74]

Interestingly, these population differences are not limited to large, diverse regions but can also be observed in smaller, isolated areas, such as the Azores islands and Sardinia, reinforcing the importance of this concept. Despite being part of Portugal, the genetic profile of the Azorean population differs significantly from that of mainland Portugal, indicating that it should be considered a distinct population for pharmacogenomic considerations.[75] Similarly, the Sardinian population has a distinct genetic background compared to the broader Italian population influencing the response to drugs and thus, a different genome-based therapeutic intervention should be followed. These two examples highlight the importance of considering genetic diversity and interbreeding over strict geographical boundaries when stratifying individuals for pharmacogenomic applications.[75,76]

In conclusion, most pharmacogenetic studies have demonstrated that the *CYP2D6, CYP2C19, DPYD, TPMT, NUDT15, SLC22A1, CFTR, HLA-A, HLA-B*, and *G6PD* pharmacogenes exhibit significant genetic differences both within and between continents. These findings underscore the need for further assessment in the near future.[54,77-80] However, it has extensively been reported that in these studies the criteria referring to the grouping of populations should be carefully considered as they might lead to inaccurate conclusions or misinterpretations of genetic variations for a corresponding population. In fact, Yang et al.[81], along with similar research, have emphasized that genetic ancestry plays a crucial role in population pharmacogenomics and strongly recommend its consideration for effective population stratification.[81,82]

Advancing population pharmacogenomics: Clinical impact and benefits

Population pharmacogenomics has the potential to influence clinical practices and provide benefits for personalized medicine, especially in developing countries. It is well recognized that pharmacogenomic analysis is not universally accessible due to the high costs and limited availability of advanced genomic technologies and resources, as previously mentioned. Ongoing efforts aim to reduce the costs of sequencing (see Chapter 5); however, these advancements have not yet brought prices to an affordable level for developing countries. Consequently, the use of advanced computational methods, tools, and algorithms to analyze and interpret population sequencing data from diverse populations can ensure drug therapy are taiolred not only to individuals but also to genetically distinct populations for commonly prescribed drugs. This approach could lead to the development of population-specific panels that can be utilized before pharmacogenomic testing, where applicable.[24,65,83] Although genomic data is often more limited in developing countries compared to developed nations, population sequencing data from publicly available databases may overcome these issues.

In addition, extensive reports reveal discrepancies between regulatory bodies regarding the available guidelines and recommendations for specific drugs, highlighting the need for these guidelines to be carefully considered and tailored to specific individuals and populations of interest. In most clinical trials, the majority of participants are recruited from populations in Europe and the USA, raising concerns about the universal applicability of pharmacogenetic guidelines. Indicatively, in U.S.-based clinical trials, approximately 84.2% of participants are of European ancestry, while only 8% are individuals of African American descent. Meanwhile, Asian individuals and Native Hawaiian/Pacific Islanders are estimated to encompass 3.4% and 0.1% of the cohort, respectively. This directly impacts the issue of inclusivity in clinical trials, which may limit the effectiveness of guidelines for diverse global populations.[84] Hence, the publicly available pharmacogenetic guidelines, which are primarily based on data from specific populations, do not accurately reflect global diversity, leading to significant limitations in their widespread implementation. To address this, it is essential to develop population-specific pharmacogenetic guidelines and recommendations. By considering the genetic diversity within specific regions, we can have significant public health benefits by identifying population-specific genetic risk factors and tailoring therapeutic interventions accordingly.

Recent advances in bioinformatics, machine learning, and artificial intelligence allow us to collect and analyze complex genetic data, identify patterns, and predict an outcome with greater accuracy.[85] This can facilitate the development of more precise and personalized treatment strategies for everyone, offering unique insights into genetic diversity and the identification of rare and novel genetic variants in developing countries, contributing to global genomic research.[61]

To summarize, a stratified medicine approach is considered necessary and can essentially contribute either to small geographic areas (islands and indigenous regions) sharing the same ethnic group. On the other hand, it can also be useful but more complex in larger diverse regions such as America, Africa, and Asia where in most cases, they do not share similar racial and ethnic backgrounds and thus, genetic heterogeneity is expected to be higher compared to other regions.[69,70,74,86] The implementation of pharmacogenomics in developing countries presents significant opportunities and challenges while population pharmacogenomics holds promise for reshaping personalized medicine even in resource-limited settings and conditions.

Drug-gene interactions in cardiology, psychiatry, and oncology

In addition to the influence of population differences on pharmacogenes relevant to commonly pre-scribed medications, distinct patterns have emerged within specific drug categories across various populations.[49,87] Notably, some populations exhibit a lower risk profile for certain drugs within categories such as cardiovascular (i.e., statins and angiotensin drugs) or antidepressant medications (i.e., SSRIs and SNRIs) compared to others. The risk proximity of drug response or toxicity can vary significantly depending on the genetic variants present within different populations.[71,88] Genetic variants that may elevate the risk of ADRs in one population might have little to no effect, or may even be absent, in another. Recognizing these variations is crucial for personalizing treatments to individual genetic profiles, thereby improving drug efficacy and reducing adverse effects, particularly in diverse populations.[65,89,90] This underscores the critical need to investigate drug-gene interactions within individual populations to optimize drug treatments. Such patterns may reflect evolutionary processes or adaptations to specific diseases within those populations.[91-93] Over time, populations may have developed genetic variations that confer advantages or disadvantages in their response to particular drugs or diseases. In the subsequent paragraphs, we will present case studies from three major areas of pharmacogenomics: cardiology, psychiatry, and oncology. These case studies will illustrate the importance of incorporating population pharmacogenomics into the framework of stratified medicine.

In cardiology, pharmacogenomic insights are essential for managing conditions such as hyper-tension, heart failure, and arrhythmias. Cardiovascular diseases (CVD) are the leading cause of death worldwide, accounting for approximately 17.9 million deaths annually, according to the WHO. Despite rapid advancements in drug development for CVD, addressing this issue remains chal-lenging.[94] Adverse drug events are common in heart conditions, influenced by factors such as disease-related alterations in drug metabolism, polypharmacy, and advanced age.[93-95] Key genetic loci involved in drug toxicity include *SLCO1B1*, *VKORC1*, and *CYP2C9* genes, which are associated with the metabolism of drugs such as HMG-CoA reductase inhibitors (statins) and anticoagulants. For instance, genetic variations in *CYP2C9* and *VKORC1* significantly impact warfarin metabolism, a widely used anticoagulant. These variations can lead to varying drug responses among individuals of different ethnic backgrounds, affecting both the safety and efficacy of treatment.[93,96] African Americans, for example, often require higher doses of warfarin due to less frequent variants such as *CYP2C9*2* (rs1799853, 430C > T) and *CYP2C9*3* (rs1057910, 1075A > C), which reduce drug metabolism, compared to Europeans.[72,97] Similarly, beta-blockers and statins, which are commonly prescribed for cardiovascular conditions, exhibit varying effectiveness and side effect profiles across different populations due to genetic diversity in genes such as *ADRB1* and *SLCO1B1*, including star alleles such as *SLCO1B1*5* and *SLCO1B1*15*.[97]

In psychiatry, determining the most effective medication often involves a trial-and-error process due to the complexity of the disease. Genetic variations in drug-metabolizing enzymes such as CYP2D6 and CYP2C19, which process many antidepressants and antipsychotics, can cause substantial differences in drug plasma levels and responses. Characteristically, *CYP2C19* and *CYP2D6* genes are involved in the metabolism of approximately 30% of all medications, making any genetic variations in these enzymes particularly impactful on drug efficacy and safety.[54,91,98] Based on the latest findings, Asian populations are more frequently classified as poor metabolizers of CYP2D6 substrates compared to other populations. As a result, they may experience elevated drug concentrations and an increased risk of adverse effects when using certain atypical antipsychotics and antidepressants, including SSRIs and SNRIs.[30] This underscores the need for dose adjustments and careful monitoring to minimize adverse reactions and optimize therapeutic outcomes. Specific star alleles, such as *CYP2D6*10*, are notably more common in East Asian populations, while alleles such as *CYP2C19*3* and *CYP2C19*4* are linked to increased drug toxicity across a variety of medications.[77] Thus the prevalence of certain genetic variations can significantly influence drug response and potential side effects, highlighting the importance of personalized medicine in psychiatric care.

Last but not least, significant differences among populations have been observed in the response to various antineoplastic agents and anticancer drugs. Variations in the *TPMT* and *DPYD* genes have been extensively linked to factors that predispose individuals to drug toxicity.[99] In fact, the *TPMT* gene has been well-studied and in particular, the variant star alleles such as the *TPMT*3A*, *TPMT*3B*, and *TPMT*3C* drawing significant interest due to their evolutionary implications and prevalence across different continents.[77,100] Nevertheless, *TPMT*2* and *TPMT*16* are also important in understanding TPMT function, contributing to a heterogeneous landscape with clinical significance. For instance, TPMT catalyzes the S-methylation of thiopurine drugs such as mercaptopurine, thioguanine, and azathioprine, which are used to treat various conditions, including acute lymphoblastic leukemia and autoimmune diseases.[101] Deficiency or absence of enzyme activity due to mutations in the *TPMT* gene can lead to high concentrations of these drugs, resulting in thiopurine toxicity.[102] Regarding the distribution of key *TPMT* genotypes, it has been observed that *TPMT*3A*, a loss-of-function allele, is more prevalent among populations of European and Latino descent, as well as among Ashkenazi Jews, with a minor allele frequency of approximately 3%—4%. In contrast, *TPMT*3C* is more common in individuals of African descent, with a minor allele frequency exceeding 4%.[77,78]

Pertaining to anticancer drugs, the *DPYD* gene serves as another important pharmacogenomic biomarker, particularly linked to an increased risk of fluoropyrimidine chemotherapy-related toxicity and, more importantly, mortality.[40,92] Variants within the *DPYD* gene are associated with DPD enzyme deficiency, which can lead to severe drug-related toxic events. Unlike *TPMT* gene, recent studies have revealed greater complexity in the functional variability of *DPYD*, suggesting that a population-based survey would be highly beneficial.[77] Important star alleles responsible for clinically significant ADRs, particularly drug toxicity, include *DPYD*6* (rs1801160), *DPYD*13* (rs55886062), as well as other variants such as rs2297595 (496A > G, p.M166V) and rs17376848 (475992T > C). At the population level, variants such as *DPYD*6*, *DPYD*13*, and 496A > G, p.M166V are most prevalent in South Asia, with a frequency of approximately 8%−9%, which correlates with a higher risk of toxicity from fluoropyrimidine drugs such as capecitabine and fluorouracil.[78]

To summarize, the significance of pharmacogenomics across various fields is profound, particularly when accounting for genetic diversity within and between populations. Ethnic differences in the allele frequencies of pharmacogenes underscore the need underscore for a more personalized approach to prescribing medications.[48,103] Beyond the areas already discussed, population differences in pharmacogenomics have also been observed with drugs such as immunosuppressants, analgesics, anesthetics, and others, further emphasizing the importance of understanding drug-gene interactions across diverse groups.[11] Implementing pharmacogenomic testing in clinical practice can lead to more personalized and effective treatments, reducing the incidence of ADRs and ultimately improving patient outcomes. This approach is crucial in diverse populations where genetic variability significantly impacts drug response and other related phenotypes.[24,104] As research in population pharmacogenomics advances, it holds the potential to transform medicine by making it more inclusive and precise, ensuring that the right drug is delivered to the right population.

List of abbreviaitons

ACE Angiotensin-converting enzyme
ADME Absorption, distribution, metabolism, and excretion
ADRs Adverse drug reactions
CDC Centers for Disease Control and Prevention
CNVs Copy number variations
CPIC Clinical Pharmacogenetics Implementation Consortium
CYP450 Cytochrome P450 enzymes
DPD Dihydropyrimidine dehydrogenase
DPWG Dutch Pharmacogenetics Working Group
EMA European Medicines Agency
FDA U.S. Food and Drug Administration
GPCRs G-protein coupled receptors
MAF Minor allele frequency
NGS Next-generation sequencing
PharmGKB Pharmacogenomics Knowledge Base
PGx Pharmacogenomics
SLC Solute carrier transporters
SNRIs Serotonin-Norepinephrine Reuptake Inhibitors
SNPs Single nucleotide polymorphisms
SNVs Single nucleotide variants
SSRIs Selective serotonin reuptake inhibitors
WHO World Health Organization

References

1. Hassan M, Awan FM, Naz A, et al. Innovations in genomics and big data analytics for personalized medicine and health care: a review. *Int J Mol Sci.* 2022;23. https://doi.org/10.3390/IJMS23094645.
2. Jørgensen JT. Twenty years with personalized medicine: past, present, and future of individualized pharmacotherapy. *Oncologist.* 2019;24:e432. https://doi.org/10.1634/THEONCOLOGIST.2019-0054.
3. Pulciani S, Di Lonardo A, Fagnani C, Taruscio D. P4 medicine versus Hippocrates. *Ann Ist Super Sanita.* 2017;53:185−191. https://doi.org/10.4415/ANN_17_03_02.
4. Müller Dj DR. *Psychiatric Pharmacogenetics: Concepts and Cases.* Cambridge University Press; 2016 (n.d.).
5. Abrahams E, Silver M. Personalized medicine for diabetes: the case for personalized medicine. *J Diabetes Sci Technol.* 2009;3:680. https://doi.org/10.1177/193229680900300411.
6. Midyett LK. One size fits all versus individualized medicine in type 1 diabetes management. *Diabetes Technol Therapeut.* 2023;25:S42−S47. https://doi.org/10.1089/DIA.2023.0109.
7. Bochud M, Guessous I. Gene-environment interactions of selected pharmacogenes in arterial hypertension. *Expet Rev Clin Pharmacol.* 2012;5:677−686. https://doi.org/10.1586/ECP.12.58.
8. Karamperis K, Wadge S, Koromina M, Patrinos GP. Genetic testing. *Applied Genom Pub Health.* 2020: 189−207. https://doi.org/10.1016/B978-0-12-813695-9.00010-8.
9. Bielinski SJ, St Sauver JL, Olson JE, et al. Cohort profile: the right drug, right dose, right time: using genomic data to individualize treatment protocol (right protocol). *Int J Epidemiol.* 2020;49:23. https://doi.org/10.1093/IJE/DYZ123.
10. Aneesh TP, Sekhar MS, Jose A, Chandran L, Zachariah SM. Pharmacogenomics: the right drug to the right person. *J Clin Med Res.* 2009;1:191. https://doi.org/10.4021/JOCMR2009.08.1255.
11. Katara P, Yadav A. Pharmacogenes (PGx-genes): current understanding and future directions. *Gene.* 2019; 718. https://doi.org/10.1016/J.GENE.2019.144050.
12. Cerpa LC, Varela NM, Martínez MM, Quiñones LA. Pharmacogenomics: basis and milestones. *ADME Encyclopedia.* 2022:874−883. https://doi.org/10.1007/978-3-030-84860-6_124.
13. Goh LL, Lim CW, Sim WC, Toh LX, Leong KP. Analysis of genetic variation in CYP450 genes for clinical implementation. *PLoS One.* 2017;12. https://doi.org/10.1371/JOURNAL.PONE.0169233.
14. Ahmed S, Zhou Z, Zhou J, Chen SQ. Pharmacogenomics of drug metabolizing enzymes and transporters: relevance to precision medicine. *Dev Reprod Biol.* 2016;14:298. https://doi.org/10.1016/J.GPB.2016.03.008.
15. Liu X. Transporter-mediated drug-drug interactions and their significance. *Adv Exp Med Biol.* 2019;1141: 241−291. https://doi.org/10.1007/978-981-13-7647-4_5.
16. Giacomini KM, Huang SM, Tweedie DJ, et al. Membrane transporters in drug development. *Nat Rev Drug Discov.* 2010;9(3):215−236. https://doi.org/10.1038/nrd3028, 9 (2010).
17. Arbitrio M, Di Martino MT, Scionti F, Barbieri V, Pensabene L, Tagliaferri P. Pharmacogenomic profiling of ADME gene variants: current challenges and validation perspectives. *High Throughput.* 2018;7. https://doi.org/10.3390/HT7040040.
18. Sultana J, Cutroneo P, Trifirò G. Clinical and economic burden of adverse drug reactions. *J Pharmacol Pharmacother.* 2013;4:S73. https://doi.org/10.4103/0976-500X.120957.
19. Zanger UM, Schwab M. Cytochrome P450 enzymes in drug metabolism: regulation of gene expression, enzyme activities, and impact of genetic variation. *Pharmacol Ther.* 2013;138:103−141. https://doi.org/10.1016/J.PHARMTHERA.2012.12.007.
20. Fischer A, Smieško M. A conserved allosteric site on drug-metabolizing CYPs: a systematic computational assessment. *Int J Mol Sci.* 2021;22. https://doi.org/10.3390/IJMS222413215.
21. Zhao M, Ma J, Li M, et al. Cytochrome p450 enzymes and drug metabolism in humans. *Int J Mol Sci.* 2021; 22:12808. https://doi.org/10.3390/IJMS222312808/S1.

22. Weinshilboum RM, Wang L. Pharmacogenomics: precision medicine and drug response. *Mayo Clin Proc.* 2017;92:1711–1722. https://doi.org/10.1016/J.MAYOCP.2017.09.001.

23. Pirmohamed M. Pharmacogenomics: current status and future perspectives. *Nat Rev Genet.* 2023;24: 350–362. https://doi.org/10.1038/S41576-022-00572-8.

24. Ji X, Ning B, Liu J, et al. Towards population-specific pharmacogenomics in the era of next-generation sequencing. *Drug Discov Today.* 2021;26:1776–1783. https://doi.org/10.1016/J.DRUDIS.2021.04.015.

25. Russell LE, Zhou Y, Almousa AA, Sodhi JK, Nwabufo CK, Lauschke VM. Pharmacogenomics in the era of next generation sequencing — from byte to bedside. *Drug Metab Rev.* 2021;53:253–278. https://doi.org/10.1080/03602532.2021.1909613.

26. World Health Organization, WHO. who.int/. Assessed Aug 19 2024.

27. Centers for Disease Control and Prevention, CDC. cdc.gov/. Assessed Aug 14 2024.

28. Centers for Disease Control and Prevention, CDC. cdc.gov/medication-safety/data-research/facts-stats/index.html, (CDC Statfacts).

29. Mathur S, Sutton J. Personalized medicine could transform healthcare. *Biomed Rep.* 2017;7:3. https://doi.org/10.3892/BR.2017.922.

30. Karamperis K, Koromina M, Papantoniou P, et al. Economic evaluation in psychiatric pharmacogenomics: a systematic review. *Pharmacogenom J.* 2021;21:533–541. https://doi.org/10.1038/S41397-021-00249-1.

31. Marsh S, Van Rooij T. Challenges of incorporating pharmacogenomics into clinical practice. *Gastrointest Cancer Res.* 2009;3:206.

32. U.S. Food and Drug Administration, FDA. fda.gov. Assessed Aug 19 2024.

33. European Medicines Agency, EMA. ema.europa.eu. Assessed Aug 19 2024.

34. Sigman M. Introduction: personalized medicine: what is it and what are the challenges? *Fertil Steril.* 2018;109:944–945. https://doi.org/10.1016/J.FERTNSTERT.2018.04.027.

35. Kalow W. Pharmacogenomics: historical perspective and current status. *Meth Mol Biol.* 2005;311:3–15. https://doi.org/10.1385/1-59259-957-5:003.

36. Pirmohamed M. Pharmacogenetics: past, present and future. *Drug Discov Today.* 2011;16:852–861. https://doi.org/10.1016/J.DRUDIS.2011.08.006.

37. Swen J, van der Wouden CH, Manson LE, et al. A 12-gene pharmacogenetic panel to prevent adverse drug reactions: an open-label, multicentre, controlled, cluster-randomised crossover implementation study. *Lancet.* 2023;401(10375):347–356. https://doi.org/10.1016/S0140-6736(22)01841-4.

38. Thorn CF, Klein TE, Altman RB. PharmGKB: the pharmacogenomics Knowledge Base. *Methods Mol Biol.* 2013;1015:311–320. https://doi.org/10.1007/978-1-62703-435-7_20.

39. National Center for Biotechnology Information. ClinVar. ncbi.nlm.nih.gov/clinvar/. Assessed Aug 5 2024.

40. Relling MV, Klein TE. CPIC: clinical pharmacogenetics implementation Consortium of the pharmacogenomics research network. *Clin Pharmacol Ther.* 2011;89:464–467. https://doi.org/10.1038/CLPT.2010.279.

41. Pharmacogene Variation Consortium. Pharmvar. pharmvar.org. Accessed Aug 5 2024.

42. Clinical Pharmacogenetics Implementation Consortium. CPIC. cpicpgx.org. Assessed Aug 18 2024.

43. DPWG, DPWG, (n.d.). knmp.nl (accessed December 10, 2024).

44. Kalman LV, Agúndez JAG, Appell ML, et al. Pharmacogenetic allele nomenclature: international workgroup recommendations for test result reporting. *Clin Pharmacol Ther.* 2016;99:172. https://doi.org/10.1002/CPT.280.

45. Gulilat M, Lamb T, Teft WA, et al. Targeted next generation sequencing as a tool for precision medicine. *BMC Med Genom.* 2019;12. https://doi.org/10.1186/S12920-019-0527-2.

46. Siamoglou S, Karamperis K, Mitropoulou C, Patrinos GP. Costing methods as a means to measure the costs of pharmacogenomics testing. *J Appl Lab Med.* 2020;5:1005–1016. https://doi.org/10.1093/JALM/JFAA113.

47. Patrinos GP. Population pharmacogenomics: impact on public health and drug development. *Pharmacogenomics*. 2018;19:3–6. https://doi.org/10.2217/PGS-2017-0166.

48. Patrinos GP. Sketching the prevalence of pharmacogenomic biomarkers among populations for clinical pharmacogenomics. *Eur J Hum Genet*. 2020;28:1. https://doi.org/10.1038/S41431-019-0499-X.

49. Westervelt P, Cho K, Bright DR, Kisor DF. Drug–gene interactions: inherent variability in drug maintenance dose requirements. *Pharm Therapeut*. 2014;39:630. /pmc/articles/PMC4159057/.

50. Malki MA, Pearson ER. Drug–drug–gene interactions and adverse drug reactions. *Pharmacogenomics J*. 2019;20(3):355–366. https://doi.org/10.1038/s41397-019-0122-0, 20 (2019).

51. Dw N, Ag M. Pharmacogenomics, ethnicity, and susceptibility genes. *Pharmacogenom J*. 2001;1:19–22. https://doi.org/10.1038/SJ.TPJ.6500002.

52. Suarez-Kurtz G. Pharmacogenomics in admixed populations. *Trend Pharmacol Sci*. 2005;26:196–201. https://doi.org/10.1016/J.TIPS.2005.02.008.

53. Ahn E, Park T. Analysis of population-specific pharmacogenomic variants using next-generation sequencing data. *Sci Rep*. 2017;7(1):1–11. https://doi.org/10.1038/s41598-017-08468-y, 7 (2017).

54. Petrović J, Pešić V, Lauschke VM. Frequencies of clinically important CYP2C19 and CYP2D6 alleles are graded across Europe. *Eur J Hum Genet*. 2019;28(1):88–94. https://doi.org/10.1038/s41431-019-0480-8, 28 (2019).

55. Gaedigk A, Sangkuhl K, Whirl-Carrillo M, Klein T, Steven Leeder J. Prediction of CYP2D6 phenotype from genotype across world populations. *Genet Med*. 2017;19:69–76. https://doi.org/10.1038/GIM.2016.80.

56. Bamshad MJ, Wooding S, Watkins WS, Ostler CT, Batzer MA, Jorde LB. Human population genetic structure and inference of group membership. *Am J Hum Genet*. 2003;72:578. https://doi.org/10.1086/368061.

57. Goswami C, Chattopadhyay A, Chuang EY. Rare variants: data types and analysis strategies. *Ann Transl Med*. 2021;9:961. https://doi.org/10.21037/ATM-21-1635.

58. Tafazoli A, Wawrusiewicz-Kurylonek N, Posmyk R, Miltyk W. Pharmacogenomics, how to deal with different types of variants in next generation sequencing data in the personalized medicine area. *J Clin Med*. 2021;10:1–23. https://doi.org/10.3390/JCM10010034.

59. Charlesworth B. What use is population genetics? *Genetics*. 2015;200:667. https://doi.org/10.1534/GENETICS.115.178426.

60. Korfmann K, Gaggiotti OE, Fumagalli M. Deep learning in population genetics. *Genome Biol Evol*. 2023;15. https://doi.org/10.1093/GBE/EVAD008.

61. Lakiotaki K, Kanterakis A, Kartsaki E, Katsila T, Patrinos GP, Potamias G. Exploring public genomics data for population pharmacogenomics. *PLoS One*. 2017;12. https://doi.org/10.1371/JOURNAL.PONE.0182138.

62. Nagar SD, Moreno AM, Norris ET, et al. Population pharmacogenomics for precision public health in Colombia. *Front Genet*. 2019;10. https://doi.org/10.3389/FGENE.2019.00241.

63. Ramamoorthy A, Pacanowski MA, Bull J, Zhang L. Racial/ethnic differences in drug disposition and response: review of recently approved drugs. *Clin Pharmacol Ther*. 2015;97:263–273. https://doi.org/10.1002/CPT.61.

64. Nagar SD, Conley AB, Jordan IK. Population structure and pharmacogenomic risk stratification in the United States. *BMC Biol*. 2020;18:1–16. https://doi.org/10.1186/s12915-020-00875-4.

65. Kaplan JM, Fullerton SM. Polygenic risk, population structure and ongoing difficulties with race in human genetics. *Phil Trans Biol Sci*. 2022;377. https://doi.org/10.1098/RSTB.2020.0427.

66. Mersha TB, Abebe T. Self-reported race/ethnicity in the age of genomic research: its potential impact on understanding health disparities. *Hum Genom*. 2015;9:1–15. https://doi.org/10.1186/s40246-014-0023-x.

67. Mizzi C, Dalabira E, Kumuthini J, et al. A European spectrum of pharmacogenomic biomarkers: implications for clinical pharmacogenomics. *PLoS One*. 2016;11:29. https://doi.org/10.1371/JOURNAL.PONE.0162866.

68. Sahana S, Bhoyar RC, Sivadas A, et al. Pharmacogenomic landscape of Indian population using whole genomes. *Clin Transl Sci*. 2022;15:866−877. https://doi.org/10.1111/CTS.13153.

69. Rodrigues-Soares F, Peñas-Lledó EM, Tarazona-Santos E, et al. Genomic ancestry, CYP2D6, CYP2C9, and CYP2C19 among Latin Americans. *Clin Pharmacol Ther*. 2020;107:257−268. https://doi.org/10.1002/cpt.1598.

70. Naranjo MEG, Rodrigues-Soares F, Peñas-Lledó EM, et al. Interethnic variability in CYP2D6, CYP2C9, and CYP2C19 genes and predicted drug metabolism phenotypes among 6060 ibero- and native Americans: RIBEF-CEIBA Consortium report on population pharmacogenomics. *OMICS*. 2018;22:575−588. https://doi.org/10.1089/OMI.2018.0114.

71. Asiimwe IG, Pirmohamed M. Drug-drug-gene interactions in cardiovascular medicine. *Pharmgenom Pers Med*. 2022;15:879−911. https://doi.org/10.2147/PGPM.S338601.

72. De T, Sehwan Park C, Perera MA. Cardiovascular pharmacogenomics: does it matter if you're black or white? *Annu Rev Pharmacol Toxicol*. 2019;59:577. https://doi.org/10.1146/ANNUREV-PHARMTOX-010818-021154.

73. Bahar MA, Setiawan D, Hak E, Wilffert B. Pharmacogenetics of drug-drug interaction and drug-drug-gene interaction: a systematic review on CYP2C9, CYP2C19 and CYP2D6. *Pharmacogenomics*. 2017;18:701−739. https://doi.org/10.2217/PGS-2017-0194.

74. Jithesh PV, Abuhaliqa M, Syed N, et al. A population study of clinically actionable genetic variation affecting drug response from the Middle East. *Npj Genom Med*. 2022;7(1):1−11. https://doi.org/10.1038/s41525-022-00281-5, 7 (2022).

75. Branco CC, Bento MS, Gomes CT, Cabral R, Pacheco PR, Mota-Vieira L. Azores Islands: genetic origin, gene flow and diversity pattern. *Ann Hum Biol*. 2008;35:65−74. https://doi.org/10.1080/03014460701793782.

76. Idda ML, Zoledziewska M, Urru SAM, et al. Genetic variation among pharmacogenes in the Sardinian population. *Int J Mol Sci*. 2022;23. https://doi.org/10.3390/IJMS231710058.

77. Zhou Y, Lauschke VM. Population pharmacogenomics: an update on ethnogeographic differences and opportunities for precision public health. *Hum Genet*. 2022;141:1113−1136. https://doi.org/10.1007/S00439-021-02385-X.

78. Zhou Y, Dagli Hernandez C, Lauschke VM. Population-scale predictions of DPD and TPMT phenotypes using a quantitative pharmacogene-specific ensemble classifier. *Br J Cancer*. 2020;123:1782−1789. https://doi.org/10.1038/S41416-020-01084-0.

79. Wright GEB, Carleton B, Hayden MR, Ross CJD. The global spectrum of protein-coding pharmacogenomic diversity. *Pharmacogenom J*. 2018;18(1):187−195. https://doi.org/10.1038/tpj.2016.77, 18 (2016).

80. Zhang B, Lauschke VM. Genetic variability and population diversity of the human SLCO (OATP) transporter family. *Pharmacol Res*. 2019;139:550−559. https://doi.org/10.1016/J.PHRS.2018.10.017.

81. Yang HC, Chen CW, Lin YT, Chu SK. Genetic ancestry plays a central role in population pharmacogenomics. *Commun Biol*. 2021;4. https://doi.org/10.1038/S42003-021-01681-6.

82. Karamperis K, Katz S, Melograna F, Ganau FB, Van Steen K, Patrinos GP, Lao O. Genetic ancestry in population pharmacogenomics unravels distinct geographical patterns related to drug toxicity. *IScience*. 2024;27(10):110916. https://doi.org/10.1016/j.isci.2024.110916.

83. Cross B, Turner R, Pirmohamed M. Polygenic risk scores: an overview from bench to bedside for personalised medicine. *Front Genet*. 2022;13. https://doi.org/10.3389/fgene.2022.1000667.

84. Turner BE, Steinberg JR, Weeks BT, Rodriguez F, Cullen MR. Race/ethnicity reporting and representation in US clinical trials: a cohort study. *Lancet Region Health - Ameri*. 2022;11:100252. https://doi.org/10.1016/J.LANA.2022.100252.

85. Quazi S. Artificial intelligence and machine learning in precision and genomic medicine. *Med Oncol.* 2022; 39:120. https://doi.org/10.1007/S12032-022-01711-1.

86. Soko ND, Muyambo S, Dandara MTL, et al. Towards evidence-based implementation of pharmacogenomics in Southern Africa: comorbidities and polypharmacy profiles across diseases. *J Personalized Med.* 2023;13:1185. https://doi.org/10.3390/JPM13081185, 13 (2023) 1185.

87. Malki MA, Pearson ER. Drug-drug-gene interactions and adverse drug reactions. *Pharmacogenom J.* 2020; 20:355−366. https://doi.org/10.1038/S41397-019-0122-0.

88. Oslin DW, Lynch KG, Shih MC, et al. Effect of pharmacogenomic testing for drug-gene interactions on medication selection and remission of symptoms in major depressive disorder: the PRIME care randomized clinical trial. *JAMA.* 2022;328:151−161. https://doi.org/10.1001/JAMA.2022.9805.

89. Wilke RA, Lin DW, Roden DM, et al. Identifying genetic risk factors for serious adverse drug reactions: current progress and challenges. *Nat Rev Drug Discov.* 2007;6:904. https://doi.org/10.1038/NRD2423.

90. Kido T, Sikora-Wohlfeld W, Kawashima M, et al. Are minor alleles more likely to be risk alleles? *BMC Med Genom.* 2018;11. https://doi.org/10.1186/S12920-018-0322-5.

91. Skokou M, Karamperis K, Koufaki MI, et al. Clinical implementation of preemptive pharmacogenomics in psychiatry. *EBioMedicine.* 2024;101. https://doi.org/10.1016/J.EBIOM.2024.105009.

92. Sharma BB, Rai K, Blunt H, Zhao W, Tosteson TD, Brooks GA. Pathogenic DPYD variants and treatment-related mortality in patients receiving fluoropyrimidine chemotherapy: a systematic review and meta-analysis. *Oncol.* 2021;26:1008−1016. https://doi.org/10.1002/ONCO.13967.

93. Dávila-Fajardo CL, Díaz-Villamarín X, Antúnez-Rodríguez A, et al. Pharmacogenetics in the treatment of cardiovascular diseases and its current progress regarding implementation in the clinical routine. *Genes.* 2019;10. https://doi.org/10.3390/GENES10040261.

94. Koufaki MI, Fragoulakis V, Díaz-Villamarín X, et al. Economic evaluation of pharmacogenomic-guided antiplatelet treatment in Spanish patients suffering from acute coronary syndrome participating in the U-PGx PREPARE study. *Hum Genom.* 2023;17. https://doi.org/10.1186/S40246-023-00495-3.

95. Faulx MD, Francis GS. Adverse drug reactions in patients with cardiovascular disease. *Curr Probl Cardiol.* 2008;33:703−768. https://doi.org/10.1016/J.CPCARDIOL.2008.08.002.

96. Božina N, Kirhmajer MV, Šimičević L, et al. Use of pharmacogenomics in elderly patients treated for cardiovascular diseases. *Croat Med J.* 2020;61:147. https://doi.org/10.3325/CMJ.2020.61.147.

97. Limdi NA, Arnett DK, Goldstein JA, et al. Influence of CYP2C9 and VKORC1 on warfarin dose, anticoagulation attainment and maintenance among European American and African Americans. *Pharmacogenomics.* 2008;9:511. https://doi.org/10.2217/14622416.9.5.511.

98. Austin-Zimmerman I, Wronska M, Wang B, et al. The influence of CYP2D6 and CYP2C19 genetic variation on diabetes mellitus risk in people taking antidepressants and antipsychotics. *Genes.* 2021;12. https://doi.org/10.3390/GENES12111758/S1.

99. Franczyk B, Rysz J, Gluba-Brzózka A. Pharmacogenetics of drugs used in the treatment of cancers. *Genes.* 2022;13. https://doi.org/10.3390/GENES13020311.

100. Ameyaw MM, Collie-Duguid ESR, Powrie RH, Ofori-Adjei D, McLeod HL. Thiopurine methyltransferase alleles in British and Ghanaian populations. *Hum Mol Genet.* 1999;8:367−370. https://doi.org/10.1093/HMG/8.2.367.

101. Pratt VM, Cavallari LH, Fulmer ML, et al. TPMT and NUDT15 genotyping recommendations: a joint consensus recommendation of the association for molecular pathology, clinical pharmacogenetics implementation Consortium, College of American Pathologists, Dutch pharmacogenetics working group of the Royal Dutch Pharmacists association, European Society for pharmacogenomics and personalized therapy, and pharmacogenomics knowledgebase. *J Mol Diagn.* 2022;24:1051. https://doi.org/10.1016/J.JMOLDX.2022.06.007.

102. Franca R, Zudeh G, Pagarin S, et al. Pharmacogenetics of thiopurines. *Cancer Drug Resist.* 2019;2:256. https://doi.org/10.20517/CDR.2019.004.
103. Lu YF, Goldstein DB, Angrist M, Cavalleri G. Personalized medicine and human genetic diversity. *Cold Spring Harb Perspect Med.* 2014;4. https://doi.org/10.1101/CSHPERSPECT.A008581.
104. Bachtiar M, Lee CGL. Genetics of population differences in drug response, current genetic. *Med Rep.* 2013; 1(3):162−170. https://doi.org/10.1007/S40142-013-0017-3, 1 (2013).

Direct-to-consumer (DTC) genetic testing and the population genomics industry

Caitlin Uren[1] and Desiree C. Petersen[2]

[1]*South African Medical Research Council Centre for Tuberculosis Research, Division of Molecular Biology and Human Genetics, Department of Biomedical Sciences, Stellenbosch University, Stellenbosch, South Africa;* [2]*South African Medical Research Council Centre for Tuberculosis Research, Division of Molecular Biology and Human Genetics, Faculty of Medicine and Health Sciences, Stellenbosch University, Cape Town, South Africa*

What is direct-to-consumer (DTC) testing?

Generally, genetic testing is offered by healthcare specialists, providing guidance with the selection of a specific test and thereafter offering assistance with the interpretation of the results obtained from screening a patient's DNA sample. Moreover, this type of testing is referred to as clinical genetic testing, which is routinely used in order to confirm a disease diagnosis. Direct-to-consumer (DTC) testing is however very different as it involves the direct selling of genetic tests to a customer via various marketing and/or online platforms. It essentially involves the advertisement of genetic test kits that are delivered directly to the interested individual's address for the self-collection of a DNA sample before returning the kit to the service provider for processing. There is not necessarily a healthcare specialist involved in the process, which means that the results are directly sent to the customer for their own interpretation. As technology has advanced and become more cost-effective, the general public has gained increased access to these DTC testing platforms. In the health and clinical fields to date, the most pronounced technology advances have been in the field of human genetics.

Various companies are offering a range of DTC genetic tests, including ancestry, common trait associations, linking diet and health aspects, as well as pharmacogenetic analysis for individual-specific drug responses. In recent years, the number of service providers is increasing in the developing world, with the regulation and ethical considerations surrounding this type of testing remaining quite limited. It largely falls on the customer to determine the suitability and value of having the test done before proceeding with these DTC genetic services.

Over the past 20 years, there have been numerous large-scale projects aimed at sequencing human populations, characterizing genetic diversity, and linking this to various phenotypes, including diseases and traits. This information informs the specific content, type of testing and conclusions made from DTC tests. Therefore, one can reliably argue that the accuracy of DTC tests relies on the extent of genomic information and associated studies that are available. Many of these studies, albeit those on

Table 9.1 Genomic projects over the years.

Project	Geographical location of individuals included in project	Developing countries included?
Human Genome Diversity Project[1]	World-wide	Yes
Simons Genome Diversity Project[2]	World-wide	Yes
1000 Genomes[3]	World-wide	Yes
African Genome Variation Project[4]	Africa only	Yes
Consortium on Asthma among African-ancestry Populations in the Americas[5,6]	World-wide	Yes
Human, Health, and Heredity in Africa[7]	Africa only	Yes
Population Architecture in Genomics and Epidemiology[8]	World-wide	Yes
UK Biobank[9]	World-wide	No
Singapore Sequencing Malay Project[10]	Asia only	Yes
Singapore Sequencing Indian Project[11]	Asia only	Yes
Singapore Genome Variation Project[12]	Asia only	Yes
MX BioBank Project	Latin America only	Yes
Oceanian Genome Variation Project	Oceania only	Yes
Chile Genómico	Chile	Yes
AfroMex Genomics Project	Mexico only	Yes
Gambian Genome Variation Project	Gambia only	Yes

the smaller scale, encompass genomic data of individuals in developing countries. These projects are summarized in Table 9.1 below.

One conclusion that can be drawn from the table above is that there are a number of projects that have genetic data on individuals in developing countries, however, this data has not been generated within or analyzed by scientists performing research in these specific countries. Another conclusion, when looking at the sample sizes of the studies above in developing versus developed countries, we can see that there is far more genetic data of populations in developed countries than in developing countries. With the availability of genomic technology increasing in developing countries, it is clear that we are coming closer to bridging this gap and introducing DTC testing more routinely in developing countries.

What are the logistics behind a DTC test?

A typical procedure to order a test, collect and return the sample and to receive the results and genetic data is presented in Fig. 9.1 below. A DTC test is routinely ordered online through the service providers website and a sample collection kit is shipped to the client's address of choice. The sample collection method can include saliva, buccal swab, blood spot or a small skin sample. The sample is then shipped back to the DTC service provider for analysis in a laboratory. The rest of the process, including the reporting of data can take anywhere from 1–12 weeks. Once the genetic data and associated report is generated, this is sent through an email, Cloud based storage link or similar. Only in rare cases will a hard copy report be generated and sent to customers. In addition, in most cases, the

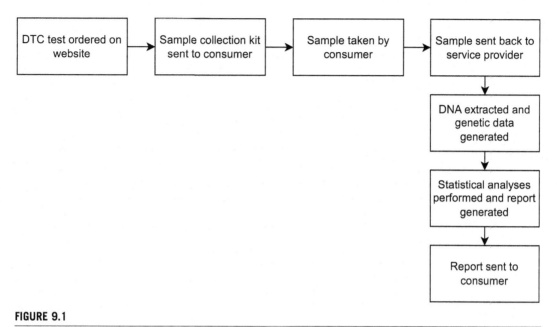

FIGURE 9.1

Logistics of a DTC genetic test.

consumer is able to download their raw genetic data for secondary analyses using different service providers or online portals established by other companies for the upload of data files to obtain additional genetic reports.

DTC genetic tests currently available

Many companies offering DTC testing will highlight the benefits of their specific test, leaving the customer with the task of having to determine which test would be most informative and assist with providing the results they are seeking. There are two main categories of DTC testing, non-health-related and health-related. In this section, we will discuss currently available DTC tests under each category, provide a brief history of the DTC test, address its availability world-wide, and where applicable, any caveats.

Nonhealth related DTC tests

Nonhealth related DTC tests largely include those relating to ancestry, genealogy, and novelty genetic traits.

   Ancestry DTC tests aim to quantify and characterize genetic diversity and ancestry contributions across the genome. In this way, global ancestry proportions and in some cases local ancestry traits are reported on. In addition, many ancestry tests screen for markers on the paternally inherited Y chromosome (in males only) and the maternally inherited mitochondrial genome (in males and females). This information therefore not only provides overall ancestral contributions but will allow for the tracking of origins with maternal and paternal lineages over generations. In the end, ancestry tests are able to detect modern admixture as well as more ancient admixture events or geographical origins of ancestral contributions.

Genealogy DTC tests aim to facilitate the tracing of family members through genetic similarity, often referred to as familial relatedness. These tests are generally performed making use of genotyping arrays containing hundreds of thousands of genetic markers resulting in only a snapshot of the autosome; the inferences made regarding relatedness based on this snapshot is however robust. Genealogy reports contain the degree of relatedness in terms of a quantifiable amount of DNA that is shared and based on this, the software's interpretation of the number of generations ago where the relation is found. If genetic data on X or Y chromosomes is available, it can also be estimated on which side of the family this relation is present.

Novelty trait DTC tests report on physical characteristics of an individual (phenotype) and generally includes the likelihood of a specific trait appearing e.g., baldness or the percentage of a trait that can be explained by genetics e.g., propensity to feel persistent coldness. While some commercial companies include this in their standard ancestry reports, the majority of the reports are from companies that accept the upload of previously generated genetic data and merely interpret this data through reanalysis. There has been criticism surrounding this interpretation and the usefulness of these tests/reports—this will be discussed in great detail in the sections that follow.

Service providers offering commonly available ancestry, genealogy and novelty trait DTC tests are detailed in Table 9.2 below. In addition, please see Case Study 1 (and) for an example of some of the DTC test reports.

Table 9.2 DTC ancestry, genealogy and novel trait tests—should we add more service providers from other developing countries?

Service provider	Interesting characteristic	Novelty traits included?	Country of supplier	Presence in developing countries?
23andMe	Largest DTC testing company	Yes	US	Yes
AncestryDNA	Large focus on research	Yes	US	Yes
MyHeritage	Focus on genealogy	Yes	Israel	Yes
FamilyTreeDNA	First DTC test company	No	US	Yes
LivingDNA	Focus on well-being	No	UK	Yes
BioCertica	Large number of different types of tests	No	South Africa	Yes
Be Happy To Be You (now called the Mediclinic Precise Ancestry Test)	First DTC test company based in Africa	No	South Africa	Yes

Health-related DTC tests

Health related DTC tests largely include those relating to the likelihood of the consumer developing a specific disease/health-related condition, the differential metabolism of medications, diet recommendations, including the use of supplements and vitamins, etc. and the tailoring of lifestyle behaviors based on an individual's genome. Most of these DTC tests report on complex phenotypes rather than Mendelian disorders (see Chapter 1 for a description of these terms). In most cases, no healthcare specialist or medical professional is involved in this DTC process both before and after receiving the final test report.

Polygenic risk score based (PRS), health predisposition, disease carrier screening DTC tests report on the likelihood of a specific health condition presenting e.g., diabetes. Similar to the novelty trait DTC tests, some companies generate this report in addition to the non-health related DTC reports, but the majority is generated from secondary analyses on other service provider websites or portals where consumers are able to upload their raw genetic data.

Service providers offering commonly available genetic predisposition DTC tests and novel characteristics of these tests are detailed in Table 9.3 below.

Pharmacogenetic (PGx) DTC tests aim to provide their customers with information on how genetic variation in their genome affects the metabolism of specific drugs, the likelihood of developing an adverse drug reaction, and provides recommendations on adjusting medication/supplement dosage in some cases. While the other DTC tests previously reported use the same genotyping array with specific genetic markers to generate the report, in the case of a PGx DTC test, there are specific arrays tailored to provide drug metabolism information.

Commonly available PGx DTC tests and novel characteristics of these tests are detailed in Table 9.4 below.

Wellness DTC tests provide information on what an individual's diet should include in order to lose weight or be healthy, dependent on their genetic variation found within their genome. Similar to the novelty trait DTC test, there has been criticism surrounding the interpretation and the usefulness of the test/report without the intervention of a healthcare specialist or medical professional—this will be discussed in great detail in the sections that follow.

Commonly available dietary DTC tests and novel characteristics of these tests are detailed in Table 9.5 below.

Table 9.3 DTC health predisposition tests.

Service provider	Interesting characteristic	Country of supplier	Presence in developing countries?
23andMe	Was one of the first companies to offer this test	US	Yes
EasyDNA	One of the cheapest offerings	World-wide	Yes
BioCertica	Includes a mental health aspect which is unique	South Africa	Yes
International Biosciences	Offers human and nonhuman tests	World-wide	Yes

Table 9.4 Service providers offering PGx DTC tests.

Service provider	Interesting characteristic	Country of supplier	Presence in developing countries?
GetMyDNA	Focus on PGx	US	No
GENEdiagnostics	Company offers whole exome sequencing too	South Africa	Yes
Be Happy To Be You (now called Mediclinic Precise Ancestry Test)	Emphasis on drugs taken to treat COVID-19	South Africa	Yes
GENETWORX	Focus on health genetic tests only	US	No

Table 9.5 Service providers offering dietary DTC tests.

Service provider	Interesting characteristic	Country of supplier	Presence in developing countries?
AncestryDNA	Can accompany the ancestry test	US	Yes
GENEWAY	Offers a wide variety of well-being tests	South Africa	Yes
LivingDNA	Presence mainly in Europe	UK	Yes
homeDNAdirect	Large variety of tests, including paternity	World-wide	Yes

Benefits of and opportunities related to DTC testing

In this section, we will be discussing the benefits of DTC testing and the unique opportunities it provides to various stakeholders, particularly when considering that some individuals residing in developing countries do not have access to all the DTC tests that are currently available. At the end of the section, we will provide a real-world example showing how some of these benefits come to pass.

Increased accessibility and convenience

Most DTC tests are available anywhere, anytime and to anyone. This means an increase in accessibility over other genetic tests. In addition, this means that consumers are able to receive their results faster as there is no need to wait for scheduled appointments with healthcare specialists. In turn, this also means that consumers do not need to visit a doctor or hospital to interpret their genetic results, decreasing the exposure to potentially infectious individuals, which was especially evident during the COVID-19 pandemic.

DTC tests are incredibly convenient as alluded to above. Adding to this, is the ability to view results on an electronic device e.g., mobile phone or computer—in this way, there is no need to wait for a hard copy report. In some instances, once the reports are received, there may also be an opportunity to make a telemedicine appointment to discuss the results. The availability of this specific service is however extremely limited for consumers based in developing countries.

Decreased cost

As briefly mentioned above, since there is no middleman, the cost for DTC tests is generally lower than that of those where clinicians or other medical professionals are involved. In addition, consumers are able to search and shop online for the lowest price offered across all service providers, and if applicable, use loyalty schemes, such as those linked to banking institutions, as well as alternative forms of payment to purchase a DTC test. This is not always an option that is available when purchasing other clinical genetic tests.

As with most products, when the general public demand increases, the price of DTC tests decreases. This can clearly be seen from the ever-decreasing cost of DTC tests since its initial offering.

Improved confidentiality

Since there is no middleman and only the service provider stores personal information, there is a decrease in the possibility of a confidentiality breach. Since the era of General Data Protection Regulation (GDPR) in Europe and Protection of Personal Information Act (POPIA) in South Africa, there is a need for service providers to only collect and process the bare minimum information collected from consumers, and there are stronger regulations on data sharing and data storage security in most countries, including those representing the developing world.

Linked to the above, since the consumer receives their results directly, without anyone else seeing or reviewing the generated report, it ensures that no-one is able to act on these results and also has no access to it. These parties could include the consumer's employer, other family members and health insurance companies.

Improved sense of connectedness, family, and community

DTC tests often allow for an increased sense of connectedness and community. Consumers are able to interact with a number of stakeholders, including the service provider themselves via social media or other platforms. This not only promotes a continued relationship between the two parties after the testing has been completed but may result in invaluable feedback that could guide future research and development of DTC tests.

Specifically, for ancestry DTC tests, it could provide the opportunity for consumers to better understand their heritage, giving a sense of belonging that could make them feel as though they are able to closely connect with their ethnicity, cultural traditions and various languages. The most significant advantage of performing genealogy DTC tests is allowing for the connection between family members previously unknown to each other and thereby facilitating the possibility of future social interactions amongst related individuals.

For health-related DTC tests, the increased connectedness allows consumers to connect with other individuals with similar health-related profiles, providing support and opportunities for discussion.

Furthermore, it also assists in establishing risk profiles for individuals with a similar genetic make-up, particularly for those that have not yet presented with a specific characteristic or trait.

More control and independence

Consumers have more control and independence during the DTC test process, including the informed consent and any information requested, from start to finish. They are able to control what test is purchased, when they will do the DTC test, when they want the sample collection kit to be delivered, how they return the kit, ability to follow-up directly with the service provider and finally, how they receive their report. Consumers are also able to request their raw data and control where they upload their data for any additional secondary analyses. In this way, the consumer is empowered with receiving their own genetic information and introduces a sense of independence, responsibility, and more importantly control over their data.

Early detection, intervention, prevention and family planning

This is particularly applicable to health-related DTC tests. The increased accessibility and availability of DTC tests promotes the early detection of genetic risk to specific health conditions that could also be inherited by future generations. This early detection therefore allows for early prevention or intervention. In the long run, this could promote increased awareness, a healthier lifestyle, a decrease in disease incidence and improved treatment outcomes that are particularly of interest in developing countries with disease-burdened healthcare settings.

Although more applicable to Mendelian disorders and more clinically orientated genetic tests, DTC test results could provide information to consumers that are wanting to consider family planning or are experiencing infertility problems.

Enlightening and educational

DTC tests have the ability to increase awareness around genomics and biomedical sciences in general. As an extended example, ancestry testing further enlightens consumers about how population genomics is informative for better understanding an individual's heritage and origins. This will not only improve the general public's overall understanding on the topic but may allow people across the world, including in developing countries, to comprehend the rationale behind this type of testing, making it available to the masses, and to more easily relate to scientists. In this way, consumers may gain appreciation for the field of genomics and for the scientists that have combined their expertise and skills to produce a DTC product that many could ultimately benefit from.

In addition, consumers are able to learn more about themselves, by unraveling the genetic information contained in their genomes, facilitating self-discovery.

For the greater good

As technology advances, it continuously improves and becomes more readily available. This fundamental concept is the same for DTC tests. As the costs for genotyping arrays and whole genome sequencing decrease, DTC tests will become more cost effective and accessible for all. This will eventually lead to increased representation of previously underrepresented populations, particularly those in developing countries. More importantly, increased representation in worldwide genomic datasets will improve DTC tests sensitivity and accuracy in the long run. This will help all consumers and service providers going forward.

Although health-related DTC tests have direct implications for the consumer, it might also have implications for family members that may not have opted to participate in this type of service offering. The same can be said for genealogy tests and finding previously unknown relatives that would prefer to remain unfound.

Adding to the potential altruistic nature of DTC tests is the ability for lay consumers to contribute their genetic data, their report, and any other information to scientists for future research. These scientists could use this data to answer specific research questions, some of which may have health implications in the long run and benefit many disease-burdened societies, often seen in developing countries.

Increased involvement and experience

DTC tests are tailor-made for the lay consumer. This means that the report is easier to read, to understand and to interpret. In this way, they are more user-friendly than other genetic test reports. In some circumstances, there is a personalized delivery of information in the form of an one-on-one session via video conference. In addition, the concept of self-discovery in the process of DTC testing is quite unique and beneficial for all.

Faster innovation and novelty

As mentioned before, with the demand for DTC tests increasing, so will the technology evolve and improve. In genomics, this could mean an increase in the number of markers genotyped or the coverage of sequencing compared to the cost, it could also mean improved accuracy in the computational steps used to generate the genotype-phenotype relationship results, or even the ancestry estimates. All in all, this will lead to an increase in innovation and novelty that is more inclusive and representative of information for many more human populations across the world.

As technology evolves, it will also become more routine for consumers to upload their raw data and use their genetic data for secondary analyses. This will ensure that the interpretation of results, is as accurate as possible. For example, requerying a variant of unknown significance might result in an improved classification of that specific variant that may be associated with a distinct trait or health condition in a previously understudied population group.

Alleviate emotional turmoil

There are a lot of emotions and possibly uncertainty that consumers experience during the DTC process. Some anxiety would be experienced, particular for health-related and genealogy DTC tests. However, once the results are received, this anxiety should dissipate if the results are expected but may cause further anxiety if they are unexpected and require lifestyle changes. Similarly, significant joy and happiness can be experienced during this process with the satisfaction that specific questions have been answered. However, if any anxiety continues, it would be recommended that individuals seek the assistance of a medical practitioner or genetic counselor for guidance and advice.

In addition, since DTC tests negate the need to visit a doctor or hospital, the anxiety caused by this process it not experienced.

Risks and controversies

In this section, we will be discussing the risks of DTC testing and the unique controversies it may lead to. At the end of the section (Box 9.1), we will provide a real-world example showing practically how some of these risks come to pass.

Privacy

One of the main issues surrounding DTC tests is privacy concerns. The majority of consumers are worried that their personal data or genetic data will become accessible to others that the data was not intended for, or that the service provider will send third parties their data without their consent. These various potential parties and the possible consequences are discussed below.

The first party that we will discuss is the consumers employer and some concerns surrounding this specific situation. If the employer gains access to their employee's genetic test results, they could discriminate against their employee if it is discovered that the individual is of a particular ancestry or related to a particular person. In addition, if it discovered that the employee has an increased risk or will go on to acquire a particular disease, the employer could dismiss the employee.

The second part that we will discuss is the consumer's family members and the general public. A concern of some consumers is that their family members either won't approve of them taking the test, won't agree with or dislike the results, or potentially, that the results may impact other family members e.g., family planning, nonpaternity etc. In addition, particularly with the uploading of personal and genetic data to websites and portals for secondary analyses, there is a concern that this data will be shared and made available to others either via website hacking or unintentionally providing the data without knowing that the data will be shared i.e., the consumer did not read the privacy policy.

The third party that we will discuss is law enforcement. If law enforcement is able to access your genetic data, they may use it to track individuals or compare to forensic samples to solve crimes. In

---

Box 9.1 Real-world stories—when DTC testing does more harm than good

Inaccurate results

There are a number of stories online regarding identical twins performing an ancestry test and getting different results. Look at this video: https://www.cbc.ca/news/science/dna-ancestry-kits-twins-marketplace-1.4980976

Data erroneously being shared with others for an unintended purpose

Whilst there are not perhaps individuals who have shared their stories of their genetic data being sent to other companies without them knowing about it, there are organizations or groups that claim that this occurs. Look at this video: https://www. wsj.com/articles/deals-give-drugmakers-rights-to-dna-data-11565607602

Data is compromised on websites where personal and genetic data was stored

Many companies' firewalls and security measures have been breached and the data downloaded without permission. Look at the following articles: https://www.buzzfeednews.com/article/peteraldhous/hackers-gedmatch-dna-privacy and https:// www.futurity.org/genetic-hacking-dna-tests-privacy-2191472-2/

some cases, the consumer's family members could be traced even though these relatives never underwent a DTC test or more importantly, consented to be traced or contacted.

The fourth party that we will discuss are health insurance companies. This is by far the greatest concern for consumers—"What will happen if my health insurance provider finds out I am likely to get cancer? Will they increase my premium or will there be certain exclusions?". These are a valid concerns and has been addressed somewhat in developed countries but is still mostly ignored in developing countries. Service providers do however need to stipulate whether the information will be shared with health insurance companies and need to make sure that the consumer consents to and understands that this will be done. If health insurance companies do however, in some way, get access to DTC test results, they may increase the consumer's premiums or deny coverage.

Lack of understanding regarding the DTC test privacy policy

As alluded to above, there is a distinct lack of understanding on the part of the consumer regarding the privacy policies of DTC test service providers. This is however normally to no fault of the consumer and rather a by-product of the way in which the policy is written, its availability on the service provider's website or the language that is used. In this way, service providers are able to inadvertently share data where in some cases, the consumer was not aware that this was taking place. Service providers should strive to make the registration and informed consent processes as clear as possible and provide a help service where consumers can easily be directed for further assistance.

Technical limitations

DTC tests are only as accurate as the science and statistics behind it. For this reason, the results of the test may not be as accurate as portrayed or clearly understood by the consumer. In addition, the sensitivity and specificity of DTC tests are highly variable—the extent of which is dependent on numerous factors, including genetic representation from different ancestries (discussed above). For example, many consumers view ancestry DTC test results as a definitive test without knowing that as more genetic data becomes available from other populations not included previously, their results may change.

In addition, for health-related DTC tests, the results may change over time. For example, a genetic marker may be associated with a particular phenotype but later this association is disproved for various reasons, including being a tag SNP for the causative genetic marker only in specific population groups, or vice versa. For this reason, interpretation of genetic data needs to be redone on more than one occasion.

Cost

DTC tests are costly and largely dependent on the type of test needed. In most cases, DTC tests are not covered by health insurance companies and will therefore need to be paid out of pocket by the consumer. In addition, confirmatory/follow-up tests may be required that will incur more costs in the long run.

For the above reasons, DTC tests are only financially accessible to those can afford to pay for it and unfortunately in developing countries, this is only a very small percentage of the population. In this way, the majority of individuals in developing countries are not able to do a DTC test and are therefore being disadvantaged by not being able to afford as well as not being represented in future DTC tests.

Clinical utility and validity

The clinical utility of DTC tests is questionable. In most cases, the marketing of service providers tends to overstate the clinical utility, suggesting that the test can predict whether the consumer will get a disease but not mentioning that there are other factors that also need to be taken into account when interpreting the DTC test results e.g., the environment and lifestyle. As a result, the lack of practically actionable information that can be gained from DTC tests results is quite low.

Similarly, the clinical validity of DTC tests remains questionable. As alluded above in the technical limitation section, the accuracy of risk determination is variable and can be low, particularly in circumstances where the statistical models behind the DTC tests have not been optimized in the population where it is supposed to be implemented. As a result, there may be an impact on false positive and false negative rates. This is particularly the case for populations in developing countries.

Risk for incorrect diagnosis

A possible false positive could result in significant and unnecessary changes in the consumer's life or the life of their family members. For example, the consumer might make lifestyle changes to improve their chances of acquiring a disease. This may result in significant costs and other negative effects as well as putting pressure on the extended family. The consumer may elect for surgeries that are unnecessary, incurring additional costs. This may also put the consumer's life at risk and increase the chance for complications.

Lastly, consumers might have misconceptions or make uninformed decisions on family planning that are unnecessary as the DTC test results are incorrect or misinterpreted by the consumer or service provider.

Overutilization and interpretation

There is a distinct overutilization of DTC tests. In some cases, particularly for the health-related DTC tests, the test is unnecessary, or the results are followed-up/confirmed with secondary genetic tests. This places a burden on service providers and healthcare resources.

As alluded to above, the results of DTC tests are often overinterpreted by consumers. Where assistance with interpretation is available, this will be an added cost, lead to inconvenience, anxiety and potentially, an unnecessary burden on the healthcare system. In most cases however, particularly in developing countries, this assistance is not readily available and would require a specific request.

Service provider-consumer relationship

The relationship between the service provider and consumer is important to keep in mind. If there is a discrepancy between genealogy records and the DTC genetic test for example, this may cause distrust by the consumer. Similarly, if there is a discrepancy between the novelty trait report and the actual phenotype for example, this may also cause distrust by the consumer and lead to further questioning the accuracy of the results.

As with most relationships, clear communication is key. If the consumer is able to understand what the test is, what it is meant for, how to interpret the results and how to get in contact with the service provider, this will all lead to a better service provider-consumer relationship in the long run as well as a better experience for all involved. This will not only benefit the consumer, but increased customer satisfaction could lead to positive marketing and in turn, an increase in sales for the service provider.

Community

During the course of the DTC test, consumers may locate unknown relatives or discover new links to other ethnicities, cultures and societies. This could lead to a questioning of self-identity with regards to which community they belong to. This self-identity may also be questioned when the interpretation of particularly, the health-related DTC test results change e.g., they thought they were at an increased risk for getting cancer, but that risk has now disappeared after they have already joined and related to cancer support communities.

Emotional turmoil

Many of the risks discussed above may lead to emotional turmoil that may also escalate over time. For example, consumers may experience anxiety waiting for or receiving the DTC test results. Consumers may also experience frustration if there is a lack of communication or support from the service provider. This may also lead to confusion regarding how to interpret the results and who to contact for assistance. Consumers may then become reluctant to perform any further DTC tests or secondary analyses and inform others of their dissatisfaction, which could be detrimental to the sustainability of the service provider.

The results of the test may make the consumer question who they are, where they belong, what they are doing in life, what changes they need to make and potentially, how the results may affect their family members. The long-term effects of these concerns are largely undocumented, more so for consumers in developing countries where other stresses such as safety and financial security take priority.

Potential for implementation in developing countries - is there a future?

As is detailed in the previous sections, there are numerous DTC tests offering various kinds of results. However, it is apparent that although service providers can ship kits to a number of countries, including developing countries, there is very little support and no local infrastructure, resources or capacity being utilized. In Africa, there are only approximately half a dozen companies that have a presence on the continent and use local infrastructure, resources and capacity.

Making use of local resources could benefit all stakeholders involved in DTC testing. Firstly, the service provider could benefit from local knowledge regarding the implementation of the test and improve the interpretation of the results given the local level of understanding regarding genomics, among other. Secondly, the consumer can benefit from a lower price and a shorter turnaround time.

Lastly, the general public can benefit from more job opportunities, increased training capacity and technological advancements.

However, the reality is that some developing countries do not have the infrastructure, resources or capacity to implement DTC testing locally. So, the question remains, what should be done to bring this to fruition? The following points should be addressed:

1. Inclusion of genomics as a subject in schools and universities
2. Inclusion of genomics, bioinformatics, biotechnology industry aspects in tertiary education facilities
3. Improved exchange programs between world-leading DTC testing service providers and scientists in developing countries interested in starting up a DTC testing company
4. Discussions with government and health insurance partners regarding funding - highlighting the dual benefit
5. Funding should be made available for start-up businesses to explore opportunities related to DTC testing.

Although infrastructure, resources and capacity are important to establish local DTC testing in developing countries, it is also vital to ensure that policies are in place to govern the industry. These policies could include the following:

1. Data governance and privacy
2. Informed consent and other ethical, legal and social implications of DTC testing

While there are policies already in place in some developing countries where local DTC testing is already implemented e.g., Protection of Personal Information Act (POPIA) in South Africa, there is no specific reference to DTC testing, the unique characteristics, and risks associated with it.

Most importantly, we need to ensure that the DTC tests implemented are scientifically valid and have been tested in local populations. It is clear, particularly for the health tests, that the genetic markers and diseases included in these service offerings should be tailored to the local population to increase sensitivity, specificity and overall interpretation accuracy. There is a definite movement toward performing this research with some continuing to develop and optimize existing DTC tests.

There is definitely a future for DTC testing in developing countries as displayed by the success of many service providers to date. Following the recommendations mentioned above and investing locally in DTC testing will bring this to fruition faster and more efficiently.

List of abbreviations

DNA	Deoxyribose Nucleic Acid
DTC	Direct-To-Consumer
GDPR	General Data Protection Regulation
PGx	Pharmacogenetics
POPIA	Protection of Personal Information Act
PRS	Polygenic Risk Score
UK	United Kingdom
US	United States

Glossary

Buccal swab A swab that captures cells from the inner cheek of the mouth

Cloud-based storage Data storage method where data is stored and transmitted on remote servers

DTC genetic testing Genetic testing that can be ordered by a consumer who then receives their results directly

Genealogy The study of tracing relatives and building family trees

Informed consent The process by which individuals provide permission knowing the full ramifications of their decision

Mendelian disorders Genetic diseases caused by single mutations occurring in a single gene

Pharmacogenetics The study of genes involved in drug metabolism

References

1. Bergström A, McCarthy SA, Hui R, et al. Insights into human genetic variation and population history from 929 diverse genomes. *Science*. 2020;367(6484). https://doi.org/10.1126/science.aay5012.
2. Mallick S, Li H, Lipson M, et al. The simons genome diversity project: 300 genomes from 142 diverse populations. *Nature*. 2016;538(7624):201−206. https://doi.org/10.1038/nature18964.
3. 1000 Genomes Project Consortium, Auton A, Brooks LD, et al. A global reference for human genetic variation. *Nature*. 2015;526(7571):68−74. https://doi.org/10.1038/nature15393.
4. Gurdasani D, Carstensen T, Tekola-Ayele F, et al. The African genome variation project shapes medical genetics in Africa. *Nature*. 2015;517(7534):327−332. https://doi.org/10.1038/nature13997.
5. Daya M, Rafaels N, Brunetti TM, et al. Association study in African-admixed populations across the Americas recapitulates asthma risk loci in non-African populations. *Nat Commun*. 2019;10(1):880. https://doi.org/10.1038/s41467-019-08469-7.
6. Mathias RA, Taub MA, Gignoux CR, et al. A continuum of admixture in the Western Hemisphere revealed by the African *Diaspora genome*. *Nat Commun*. 2016;7:12522. https://doi.org/10.1038/ncomms12522.
7. Choudhury A, Aron S, Botigué LR, et al. High-depth African genomes inform human migration and health. *Nature*. 2020;586(7831):741−748. https://doi.org/10.1038/s41586-020-2859-7.
8. Wojcik GL, Graff M, Nishimura KK, et al. Genetic analyses of diverse populations improves discovery for complex traits. *Nature*. 2019;570(7762):514−518. https://doi.org/10.1038/s41586-019-1310-4.
9. Sudlow C, Gallacher J, Allen N, et al. *PLoS Med*. 2015;12(3). https://doi.org/10.1371/journal.pmed.1001779. e1001779.
10. Wong L-P, Ong RT-H, Poh W-T, et al. *Am J Hum Genet*. 2013;92(1):52−66. https://doi.org/10.1016/j.ajhg.2012.12.005.
11. Wong L-P, Lai JK-H, Saw W-Y, et al. *PLoS Genet*. 2014;10(5). https://doi.org/10.1371/journal.pgen.1004377. e1004377.
12. Teo Y-Y, Sim X, Ong RTH, et al. *Genome Res*. 2009;19(11):2154−2162. http://genome.cshlp.org/content/19/11/2154.abstract.

Clinical translation of genomics research in Africa—Mainstreaming medical ethics, equity, genetics education, and public engagement

Nicole van der Merwe[1,2]**, Carene Anne Alene Ndong Sima**[3] **and Nchangwi Syntia Munung**[4]

[1]*Division of Chemical Pathology, Department of Pathology, Faculty of Medicine and Health Sciences, Stellenbosch University, Cape Town, South Africa;* [2]*FamGen Counselling, Bloemfontein, South Africa;* [3]*South African Medical Research Council Centre for Tuberculosis Research, Division of Molecular Biology and Human Genetics, Faculty of Medicine and Health Sciences, Stellenbosch University, Cape Town, South Africa;* [4]*Division of Human Genetics, Department of Pathology. Faculty of Health Sciences, University of Cape Town, Cape Town, South Africa*

Introduction

The global enthusiasm for genomic medicine and its associated advantages, have ignited considerable interest among African countries to initiate programs that will support and oversee its implementation. This interest is exemplified by the establishment of pilot public health genetic programs aimed at facilitating newborn screening for monogenic conditions, the creation of centers of excellence for genomic medicine in some African countries, and the implementation of training programs aimed at either improving genetic education in healthcare workers or training cohorts of genetic counselors. These different initiatives demonstrate the commitment of several African countries to harness the potential of genomic medicine and to ensure its successful integration into their healthcare systems.

As African countries prepare to embrace genomic medicine, several ethical, legal, and social issues (ELSIs) come to the fore. These include concerns around the dearth of genomic data on African populations,[1,2] feedback of individual and incidental findings,[3] community/patient involvement in decisions on secondary uses of genetic data,[4,5] the potential for stigma and discrimination,[6] and the risks of privacy breaches. In addition to these ELSIs, there are further concerns about Africa's readiness to effectively and equitably implement genomic medicine. A pressing concern is the apt shortage of a skilled genomics workforce in African countries,[7,8] which include professionals such as medical geneticists, genetic researchers, data scientists, genetic counselors, and bioethicists.[9,10]

In this chapter, we draw on academic scholarship on genomics and ethics in Africa, to provide an overview of ELSIs that may emerge in the translation of genetic research and the subsequent implementation of genomic medicine in Africa. We also highlight ongoing genomics initiatives in

Africa that are currently addressing these ELSIs. We acknowledge that some of these ELSIs are not limited to the African continent; however, our presentation of these ELSIs is largely informed by our experiences as researchers working in the field of bioethics and genetic counseling in Africa.

Informed consent comprehension and ethically appropriate consent models

Informed consent is the cornerstone for upholding the autonomy of research participants and patients. Genomics has brought about noteworthy changes in how researchers typically approach and conceptualize informed consent in health research. Generally, informed consent is study-specific. However, genomics and big data-driven research have necessitated the exploration of consent models that can accommodate secondary and third-party uses of biospecimen and data, leading to a noticeable transition from the traditional study-specific consent to alternative consent models such as broad consent, tiered consent, meta consent, and dynamic consent. Broad consent[11] allows for the storage and use of biospecimens and data for future research, irrespective of their direct relation to the initial study (a general description of potential studies however needs to be provided). It provides flexibility for researchers, enabling potential scientific advancements while respecting participant autonomy. In contrast, tiered consent gives research participants the option to specify how they would like their samples and data to be used. In addition to respecting the autonomy of research participants, tiered consent can be said to give research participants some degree of control over their samples and data. On the other hand, dynamic consent allows for continuous communication between researchers and participants, enabling participants to consent to specific uses of their data and to update their preferences over time. The transition from specific consent to these alternative models reflects the need to strike a delicate balance between advancing scientific research through secondary and third party uses of samples and data, and upholding the autonomy and rights of research participant's over the use of their samples and data.

Genomics research projects across Africa, even within the same research consortia, tend to use different consent models[12] and the health research regulations in many African countries neither fully endorse nor censure broad or tiered consent.[13] Empirical studies reveal varying perspectives on the acceptability of each of these different consent models. Researchers in Africa may favor broad consent for practical reasons related to research implementation and cost, even when they acknowledge that tiered consent is more consistent with the principle of autonomy.[14,15] Conversely, research ethics committee members across Africa have expressed reservations related to broad consent,[16,17] while research participants have little such reservations.[18,19] This diversity in stakeholder views indicate the need for normative and empirical scholarship on ethically appropriate consent models for genomics and big data—driven health research in Africa.

Beyond consent models, the comprehension of study information by research participants—a fundamental requirement of the informed consent process—remains a major challenge in many African settings. Several empirical studies have consistently reported challenges in comprehension of study information in genomic research across Africa.[20–22] Some of the factors contributing to the limited comprehension and recall of consent information include diagnostic misconception, low levels of genetic literacy, situational vulnerability, and information overload.[23,24] Diagnostic misconception refers to the phenomenon whereby individuals mistakenly perceive a research study as part of their clinical care or an opportunity to receive more information about their health condition, while situational vulnerability occurs when consent is given during emotionally distressing conditions, without an understanding of the research project.[25] Given these challenges, it is imperative to explore strategies to improve the understanding of study information in genetics research.

A study conducted in South Africa showed that iterative learning, through the use of an *assessment of capacity to consent questionnaire,* can improve research participants' understanding of consent information.[26] Similarly, a study in Nigeria showed that incorporating indigenous words and concepts to explain scientific terminology such as heritability, traits, and diseases can enhance comprehension,[27] while studies in different African countries have demonstrated that rapid ethical assessment can help simplify and contextualize consent information before the start of the study.[21,28,29] The use of one or more of these strategies should be incorporated in genomics projects in Africa to help improve comprehension of research information and to empower research participants to make informed choices regarding the use of their samples and genetic data for research and innovation.

Genetic privacy, stigmatization, and discrimination

Personal genetic information is a digital identifier that when linked to health records, can unveil deeply personal attributes not only of the individuals who provided the data, but also of their families and ethnic backgrounds or ancestral origins. This sensitive information could potentially be exploited by third parties including employers, insurance companies and schools, leading to discrimination against individuals, their families, or population groups to which they belong or are affiliated with. Thus, genetic privacy is paramount to ensure that genetic data provided as part of research or healthcare remains secure.

In most research projects, genomic data are typically kept anonymous, with the risk of re-identifying participants therefore being minimal. However, with advances in sequencing technologies, increasing accessibility of genetic testing services, and more countries preparing for digital health services, segregating genetic and phenotypic data will be counterproductive. This means that new mechanisms would have to be identified to safeguard the privacy of research participants or at the very least, minimize the risk of discrimination based on genetic information.

The enactment of data protection laws in many African countries offers some hope that discrimination based on genetic information can be mitigated. The South African legislation, for example, explicitly safeguards against discrimination and stigmatization, including genetic discrimination.[30] This is particularly important given the emergence of genetics companies in Africa with implicit objectives to generate and acquire genomic information for commercial and profit-oriented purposes. This requires establishment of robust mechanisms for data security and defining responsibilities of all stakeholders including policymakers, researchers, and healthcare professionals toward ensuring responsible and respectful use of genetic information of which they are the stewards or custodians.

For genomics to have greater impact on patient care, especially in terms of personalized medicine, there is a need to link or integrate genetic data with personal health records. This makes privacy difficult to guarantee. Several genomics research initiatives are already working towards making highly re-identifiable genomic datasets publicly available, and it has been reported that some individuals are willing to forgo genetic privacy in favor of advancing personalized medicine.[31,32] The Personal Genome Project, for example, uses an open consent model, where research participants voluntarily relinquish their genetic privacy, granting researchers permission to integrate their genetic and phenotypic information and to make the potentially identifiable information publicly available for research and scientific discovery. This shift toward sacrificing genetic privacy for the benefit research and development reflects the growing recognition of the potential benefits that can be derived from integrating genetic and health data. However, it is important to establish clear conditions under which third parties such as employers, insurance companies, and national governments can access and use personal health (both genetic and clinical) data.

Privacy protection vs. duty to warn of shared genetic risk

Genome sequencing may reveal that a patient and their first-degree biological relatives have or are at risk for a particular condition.[33] Health professionals may therefore be confronted with conflicting duties to protect a patient's privacy and to warn biological relatives of shared risk.[34] Some authorities have contended that physicians have a privilege (but no obligation) to inform relatives of possible genetic risks in cases where the patient does not voluntarily disclose risks to the family; probable serious harm is likely to occur; at-risk family members are identifiable; and the condition is preventable, treatable, or amenable to risk-reducing surveillance.[35] The recent ABC vs. St Georges Healthcare National Health Services Trust & Ors court case which took place in the High Court of Justice in England and Wales,[36] set a precedent for performing a balancing exercise by all healthcare professionals, which requires them to consider the patient's confidentiality as well as the potential duties of care to relatives. The overarching message from this case was that all healthcare professionals need to provide evidence that they have at least considered the duties of care they owe to relatives. Middleton et al.[34] contended that healthcare professionals seeing genetics patients and families must (1) clearly document in all of the patient's and relative's medical records what this "balancing exercise" has considered and what other healthcare professionals were involved in discussions around the case; (2) clearly articulate and document the arguments for or against disclosure, including an assessment of foreseeable psychological harm imposed by both courses of action; (3) be legally bound to deliver on the conclusion drawn from the balancing exercise; and (4) consider how the information should be disclosed to at-risk relatives as well as make arrangements for family members to receive genetic counseling or other support.

Defining priorities for genomics research and medicine in Africa

Over the past decade, appreciable progress has been made globally in the translation of genomics research to improve patient care. However, the adoption of genomics in clinical settings in Africa has been relatively slow, with only a handful of countries having a functional medical genetics service. In some cases, the integration of genomics into clinical practice has occurred through international research programs and public–private partnerships.[37–39] This was the case for example in South Africa, where a public–private partnership facilitated the creation of Community Genetic Services.[37] In Cameroon, an international research collaboration enabled a public hospital to pilot a medical genetic service that could provide prenatal genetic diagnosis and genetic consultations for a range of monogenic diseases,[40] arguably paving the way for the country's first medical genetic service. However, the adoption of these programs by national governments has been slow and/or encountered significant challenges and delays. This slow progress may be attributed to the idea that some of the pressing health needs of populations in Africa may not necessarily require genomics-based solutions but rather access to primary healthcare or basic services like clean water and sanitation facilities. Moreover, while the cost of genetic technologies is decreasing, the health expenditure per capita in most African countries remains significantly lower than the current cost of a DNA test.

Addressing these challenges requires a comprehensive and integrated approach involving advocating for increased global investment in genomics research in Africa and prioritizing complex diseases that have an established genetic component as well as genetic technologies with a favorable cost—benefit ratio. Examples of such priorities could include pharmacogenomics, genetic testing and screening programs such as newborn, prenatal, and premarital screening, and the identification of new therapies for common and rare monogenic conditions.[8,41] When identifying these priorities, it is crucial to engage in wider consultations with different stakeholder groups including researchers, healthcare professionals, commercial biotechnology companies, policymakers, and community representatives. It is also important to acknowledge that many African governments lack the resources or capacity to independently implement genomics research and innovation projects. Therefore, international collaborations and public—private partnerships may be essential for advancing clinical translation of genetic research in Africa. Such collaborations can help mobilize resources, expertise, and infrastructure needed to bridge the gap between genetic research and clinical applications on the African continent.

Governance of data sharing and access

The successful application of genomic medicine in African countries would to a large extent depend on the ability of local researchers and clinicians to access and analyze large volumes of genomic data. Initiatives like the Human Heredity and Health in Africa consortium (H3Africa Consortium, 2014) and the African Genome Variation Project[42] have increased the availability of data on genomic variations in African populations (see Chapter 3 for descriptions of these initiatives). Additionally, the growing presence of commercial services offering direct-to-consumer genetic testing suggests that more genetic data on African populations will become available within the next decade. Many African countries are also investing in electronic health records and patient-consented disease registries, further enhancing data availability for genomic medicine programs. Open sharing of these data for research purposes can bring numerous benefits for research translation.[43-45] However, public trust in the collection, generation, storage and re-use of genetic data and medical records will have to be guaranteed.

Several studies across Africa have explored the perspectives of different stakeholders regarding the generation and/or sharing and analysis of genomic and health data.[46-48] These studies have highlighted key concerns related to trust, the return of results, data ownership, benefits of data sharing for patients, the possibility of group stigma, inequities in collaborations, and the importance of giving research participants or data donors a say in the secondary uses of their data - whether for research, innovation, or clinical practice. For example in 2015, one of the largest private health insurance companies in South Africa announced plans to offer low cost genomic testing to their consumers.[49] The overall goal was to build a comprehensive collection of genomic, phenotypic and clinical data which their partner (Human Longevity Inc) could then sell to pharmaceutical companies. While assurances were made regarding data security, some Bioethicists[50] expressed concerns about the potential exploitation of their clients under the guise of improved clinical care. An urgent need exists to address issues related to cross-border data transfer, lack of client benefits, the return of individual results and feedback of incidental findings, as failure to address these ethical issues could erode public support and trust in the secondary use of genomic and clinical data. The above examples demonstrate the need to develop data governance models that could adequately promote transparency, account-ability, and the responsible and equitable use of genetic data for the public good.[51]

Knowledge translation: Commercialization, patents, and intellectual property rights

Translating genomic knowledge into new products and services is necessary for the transition from research to clinical care. However, this process raises complex ethical and legal issues related to commercialization and patenting. Commercialization efforts such as the one undertaken by the American genomics company and the South African health insurance company partnership[49], faced resistance due to concerns about consent, data security, benefit sharing, intellectual property rights, and patenting. A similar case was the alleged commercialization of a microarray kit developed from African DNA samples, leading to some African research institutions requesting that the samples be returned to the institutions that provided the samples as consent for commercialization was not ob-tained from them, and there were no agreements in place for benefit sharing of intellectual property rights and patenting.[52]

Patenting is one way institutions can generate financial profit from breakthroughs in scientific research and innovation. While patents can incentivize research translation efforts in genomics research,[53] there is a parallel view that gene patenting may have no important quantitative effect on scientific research and follow-on innovation.[54] Whether or not patenting benefits or restricts scientific research and innovation, the greater question is whether the benefits would be equitably distributed across nations and populations.[55] It is worth mentioning that patents and intellectual property regimes tend to work best where there is a healthy economy and are unlikely to incentivize the translation of genomics knowledge pertaining to health problems specific to Africa. The effect of this may be limited access to genomic medicine by populations in Africa, despite their participation in population ge-nomics studies.

Genomics initiatives in Africa have a responsibility to identify patent and intellectual property models that support access to innovations and benefits for the populations that provide samples and data. This involves designing mechanisms that ensure equitable distribution of benefit, as well as prioritize access to proven genomic interventions that addresses the continent's unique health chal-lenges. Additionally, stakeholders should engage in discussions and collaborations to develop ethical guidelines and policies for commercialization and intellectual property management in genomics research in Africa.

Benefit sharing

Benefit sharing is a benchmark for ethical research.[56] With origins in the field of biodiversity, the concept of benefit sharing has gradually extended to clinical research, where it was primarily understood as the obligation to provide proven interventions to research participants or study communities. In recent years, it has become a complex and contentious ethical issue in human genetics research. While a universally accepted definition of benefit sharing is lacking, a frequently cited definition is that by Doris Schroeder which describes benefit sharing as *"the action of giving a portion of advantages/profits derived from the use of human genetic resources to the resource providers to achieve justice in exchange, with a particular emphasis on the clear provision of benefits to those who may lack reasonable access to resulting healthcare products and services without providing unethical inducements."*[57] Simply put, the concept of benefit sharing centers around what is "owed" to research participants or study communities and can be viewed as a means of compensatory justice.

Studies focusing on benefit sharing in human genetics research in Africa have highlighted that expectations of benefit sharing are rooted in the principles of solidarity and reciprocity.[58] These benefits can encompass various aspects such as access to healthcare, research capacity building, access to proven interventions, technology transfer, infrastructural development, and the provision of social amenities to study communities.[58,59] Discussions on benefit sharing in genomics have mainly taken place within international academic research collaborations, and it is unclear whether similar benefits and justifications would apply to academic—industry partnerships and collaborations. Nevertheless, there is one noteworthy example of an academic—industry partnership in South Africa that has outlined a benefit sharing model. This model includes commitments from the industry partner to support community projects, share a portion of annual net revenue with community organizations and their partner networks, and, in the event of acquisition, distribute net proceeds to communities that have contributed to their genomics database. While commendable, it is essential to examine how this model addresses questions of justice and commercialization in genetics research, and whether it enables study communities to have a fairer and more active role in influencing the implementation of this benefit sharing model.

Feedback of genetic and genomic findings
Actionability in the African context

The concept of "actionability" is a universal requirement for return of genomic results. International groups and organizations have specified definitions for actionability (American College of Medical Genetics and Genomics (ACMG), ClinGen, 100,000 Genomes project), which may be extended for applicability to the African context, taking the following aspects into consideration: that it (i) makes provision for adult-onset and childhood-onset conditions which are highly prevalent in Africa — many pertinent and challenging conditions are diagnosed in children (e.g., sickle cell anemia); (ii) is not limited to highly penetrant genes but includes pharmacogenomics genes as well as high risk alleles that enable risk reduction or prevention of adverse drug reactions; and (iii) includes treatable or manageable conditions for which a specific lifestyle intervention exist (e.g., avoiding certain foods) or for which genetic counseling is possible to enable better reproductive choices.

A few studies have reported on the opinions of African stakeholders regarding the return of individual genetic results, and discussed the universal requirements for actionability and result validation in an accredited laboratory.[3,60,61] Authors acknowledged the nuances of being able to meet these requirements in the African setting in which resources are limited. In these studies, some stakeholders were of the opinion that it is important to return only results that are verified and actionable (Botswana) while others contended that all whole exome sequencing (WES) results should be confirmed with Sanger sequencing (or PCR), especially if that result is to be used for family screening or prenatal testing (South Africa). Most stakeholders in these studies contended that diagnostic validation as well as Next Generation Sequencing (NGS) data reanalysis in the African context presents a huge dilemma in terms of the infrastructure and resource availability. Even in African countries such as Botswana where healthcare is provided free of charge, this may pose a problem.[60] The fact that healthcare systems are under-resourced in most African research settings, may mean that even if participants are informed about a medically actionable finding, it may not be actionable or treatable in that particular healthcare setting, leading to the recommendation that the standard for what should be deemed actionable, should be influenced by the specific interventions that are available in a particular country, and not internationally.[6,62]

Feedback of individual genetic results and incidental findings

De Vries et al.[13] reported that only seven countries in Africa (Botswana, Cameroon, Ethiopia, Rwanda, Malawi, Sudan, and Uganda) specifically refer to the return of genetic results in their ethics regulatory frameworks for genomic research and biobanking. Uganda in particular is the only country that stipulates that any results that are of clinical relevance, including incidental findings, ought to be fed back to study participants. Return of incidental findings (IFs) in particular may evoke various emotional responses including depression, anger, guilt, or anxiety which may be worsened by a lack information on the timing or severity of the associated disease.[63,64] Beyond the consequences for a patient's psychological well-being, IFs may also affect employment, insurance coverage, and cause family discord or social stigmatization.[63-65] Families may furthermore be faced with challenging reproductive choices for a finding of carrier status for a recessively inherited condition.[65] The British Medical Association and the American Academy of Pediatrics recommended that carrier status obtained incidentally should be conveyed to parents, while the American Medical Association and the German Society of Human Genetics recommended that incidental carrier status should not be disclosed to parents or third parties but should be discussed with the child when they reach reproductive age. This is due to the potential risks including social risks (discrimination and stigmatization), psychological risks, and deprivation of reproductive privacy.[66] Recent guidelines related to carrier testing in minors are foreseeably more flexible given the advances in technology and wider use of newborn screening.[67] Some of the challenges related to the African setting include the limited number of genetics professionals and access to genetic counseling services, lack of central record-keeping facilities (files may be easily lost by the time child reaches adulthood), liberal termination of pregnancy laws in some countries, limited access to cascade testing, and a high number of teenage pregnancies.[68] Some international societies and associations haveprovided recommendations for the return of WES results, including the ACMG, the Canadian College of Medical Geneticists (CCMG), the European Society of Human Genetics (ESHG), and the American Society of Human Genetics (ASHG), among others.[69] Considerations not unique to, but paramount to the African context include additional health costs and stress for individuals and families, the availability of confirmatory testing, preventive care and ongoing screening and surveillance.

Challenges related to the above considerations which may impede the widespread implementation of WES/whole genome sequencing (WGS) in routine care include unequal access to care, resource constraints (financial, infrastructural, and workforce), and complexity of inferring variant pathogenicity due to vast genetic diversity and biased genome datasets that are skewed by the overrepresentation of individuals from well-studied groups.[70] Utilization of the wealth of genetic diversity within African populations to better characterize and understand clinical, pathological, and genetic variability is crucial. However, the current lack of available population-level genomic data renders the operationalization of the above recommendations challenging. When compared to populations from West and East Africa, genomic data for Northern, Central, and Southern African populations are particularly lacking,[42,70–72] which adversely impacts the interpretation of genomic studies in these groups.[42]

Secondary findings in the African context

In 2013, the ACMG developed recommendations for reporting a list of actionable genes when genomic sequencing is performed.[73] A similar list was subsequently developed by the 100,000 Genomes Project (www.genomicsengland.co.uk). These include genes linked to a spectrum of conditions ranging from cancer to cardiomyopathies, selected due to its association with a clear or definable set of clinical features, the possibility of early diagnosis, a reliable clinical genetic test, and effective intervention or treatment.[73] There is however limited evidence supporting the identification and proportions of pathogenic mutations in genes included in these lists, in African populations.[62]

African researchers have reported on genes associated with conditions that are highly prevalent, result in a high disease burden in certain African populations, are actionable (treatable with medication or surgery, where surveillance possible, reproductive choices exist, vaccination is available, dietary intervention is possible, contraindications exist to medical procedures and certain medications or foods) and have an adequate evidence base. These genes include *HBB* which causes sickle cell anemia and b-thalassemia due to its high disease burden, its medical actionability, and vast evidence available that enable medication prescription, surveillance, reproductive choices, and vaccination.[74–76] Another widely publicized gene variant includes *HLA-B*5701* associated with abacavir hypersensitivity re-action. This medication is the first-line treatment for HIV which is the highly prevalent in some countries (e.g., Botswana).[77–79] Another gene of interest is *G6PD*, where variations can lead to G6PD deficiency. This condition, which is prevalent among males, carries a significant disease burden in African countries with high malaria prevalence. The actionability is realized in the recommendation to avoid certain foods and medications.[80–82] The gene *APOL-1* has furthermore received much attention due to regional increases in the prevalence of kidney disease among carriers. It has been considered to be actionable due to the contraindication to kidney donation, especially since the condition is more aggressive in individuals with comorbidities and high-risk genotypes.[83–87] Health conditions in the current ACMG and UK SFs lists that seem relevant for reporting due to a high prevalence worldwide and their actionability in most countries, include familial hypercholesterolemia and certain inherited cancers with sufficient available evidence and guideline-based management recommendations. Most of the remaining genes that have been studied in African populations include those that are actionable but relatively rare, and lack sufficient data, evidence, or available treatments.

The next frontier for ethical and equity-oriented genomics in Africa
Building a critical mass of big data and genomics medicine workforce in Africa

In the near future, genomics will be an integral feature of clinical medicine in many African countries. This will have been facilitated by the increase in availability of genomic information on African populations and ongoing efforts to build capacity for genomics research in Africa. As a result, hospitals will be able to access infrastructure for genetic testing through research programs, and build staff capacity for the implementation of genomic medicine. However, many of the ethical challenges encountered in genomics research are also likely to arise in clinical practice, which may impede the implementation of genomic medicine programs. A previously discussed, dilemma is the issue of returning individual and incidental genetic findings, which remains a challenge in genetics research in Africa. Compounding this challenge is the scarcity of a medical genetics workforce who would be essential in facilitating the return of genetic results. It is estimated that one genetic counselor is required to service every 100,000 individuals of the population. However, currently, there are only ~30 practicing genetic counselors and 20 medical geneticists registered with the Health Professions Council of South Africa, which means that − considering the local burden of genetic disease and the total population size of 61 million − less than 10% of the country's need for clinical genetic services are currently being met.[10] With the increasing use of NGS technologies, the demand for skilled professionals who are able to understand and interpret complex and controversial genetic findings, will exponentially grow.[88] This demand will likely be met in two ways; by upskilling primary care health professionals including general practitioners and nurses, and upscaling current training of genetic counselors and medical geneticists.[3] With the advent of genotyping arrays and point-of-care genotyping technologies, the involvement of a genetic counselor or trained genetics professional is imperative given the medical and/or psychological implications associated with receiving immediate, complex or high-risk results, particularly in the settings of predictive and self-referred testing. Integrating genetic counselors as vital intermediaries between laboratories and clinicians, facilitates the alignment of patient expectations with guideline-based practices. As the demand for genetic counseling services increase, local medical schemes are urged to provide coverage for this specialized service and to ensure adequate compensation as is the case in first-world countries.[89] Efforts are currently underway to address the lack of trained patient-facing genetics professionals. For instance, the African Genomic Medicine Training Initiative[90] is offering short online training in genomic medicine for nurses and healthcare professionals. Moreover, some medical schools in South Africa provide training for (sub)specialist medical geneticists while the University of Ghana has recently introduced a postgraduate training program in genetic counseling. Additional programs are however needed across Africa to ensure widespread access to genomic medicine.

A social justice lens for genomics and health in Africa

Equitable access to genomic technologies remains a major concern particularly in low- and middle-income countries. Currently, the global genomic medicine landscape exhibits significant disparities, with many low- and middle-income countries lagging behind their international counterparts despite declining costs in genomic technologies. Also, existing genetics services are mainly operating in tertiary and private hospitals located in urban areas, thereby rendering it inaccessible to individuals and populations challenged by financial constraints and geographical distance. To address this challenge, genomic research initiatives, policymakers, and clinicians must explore strategies to make genomic medicine more accessible to populations across the African continent. This necessitates considering how intellectual property rights related to genomics innovation might impact equitable access to genomics technology. Additionally, identifying models for implementing benefit sharing is crucial.

These models should ensure that the benefits derived from genomic research and technological advancements are distributed fairly and reach those who need them the most.

Public engagement for genomics medicine in Africa

Addressing some of the ethical tensions/dilemmas in translational genomics would require an inclusive and participatory public engagement. Such an approach requires active listening to the diverse perspectives and lived experience of various stakeholder groups and the co-creation of governance frameworks that addresse issues of privacy, benefit sharing, ownership rights to genomic data, secondary and third-party use of genomics data, and intellectual property rights. Building public trust in genomics medicine science is essential for the successful implementation of genomic medicine in Africa. However, genetic literacy in many African countries is low, and public engagement and citizen science have not really been used in genomics and health research in Africa. Genomics research and big data in medicine initiatives will have to take proactive steps to introduce the concepts of genomics and big data in health and medicine to the general public so as to empower the public to engage meaningfully in genomic medicine initiatives.[91] This will also mean creating accessible public engagement platforms that the general public can use to share their perspectives and concerns around genomics and also contribute in shaping ethical frameworks for genomic medicine. Public engagement initiatives can include town hall meetings, workshops, online forums, and citizen science projects that empower individuals to actively participate in the ethical discourse surrounding genomics.

Acknowledgments

Nchangwi S. Munung is supported through research funding from the National Human Genome Research Institute and the Office of the Director of the National Institutes of Health (Award number U01MH127692) and the National Heart, Lung, and Blood Institute of the National Institutes of Health (Award Number: U24HL135600). The views expressed in this chapter do not necessarily represent the official views of the National Institutes of Health, and they had no role in the preparation of the manuscript nor the decision to publish it.

References

1. Ramsay M. African genomic data sharing and the struggle for equitable benefit. *Patterns*. 2022;3(1):100412. https://doi.org/10.1016/j.patter.2021.100412.
2. Tiffin N. Potential risks and solutions for sharing genome summary data from African populations. *BMC Med Genom*. 2019;12(1). https://doi.org/10.1186/s12920-019-0604-6.
3. Van Der Merwe N, Ramesar R, De Vries J. Whole exome sequencing in South Africa: stakeholder views on return of individual research results and incidental findings. *Front Genet*. 2022;13. https://doi.org/10.3389/fgene.2022.864822.
4. Munung NS, Nembaware V, de Vries J, et al. Establishing a multi-country sickle cell disease registry in Africa: ethical considerations. *Front Genet*. 2019;10. https://doi.org/10.3389/fgene.2019.00943.
5. Nordling L. Give African research participants more say in genomic data, say scientists. *Nature*. 2021; 590(7847):542. https://doi.org/10.1038/d41586-021-00400-9.
6. de Vries J, Landouré G, Wonkam A. Stigma in African genomics research: gendered blame, polygamy, ancestry and disease causal beliefs impact on the risk of harm. *Soc Sci Med*. 2020;258:113091.

7. Kamp M, Krause A, Ramsay M. Has translational genomics come of age in Africa? *Hum Mol Genet.* 2021; 30(2):R164−R173. https://doi.org/10.1093/hmg/ddab180.

8. Munung NS, Mayosi BM, de Vries J. Genomics research in Africa and its impact on global health: insights from African researchers. *Glob Health Epidemiol Genom.* 2018;3:e12. https://doi.org/10.1017/gheg.2018.3.

9. Beighton P, Fieggen K, Wonkam A, Ramesar R, Greenberg J. The University of Cape Town's contribution to medical genetics in Africa - from the past into the future. *S Afr Med J.* 2012;102(6):446−448. http://www.samj.org.za/index.php/samj/article/view/5621/4159.

10. Kromberg JG, Sizer EB, Christianson AL. Genetic services and testing in South Africa. *J Commun Gene.* 2013;4(3):413−423. https://doi.org/10.1007/s12687-012-0101-5.

11. Maloy JW, Bass PF. Understanding broad consent. *Ochsner J.* 2020;20(1):81−86. https://doi.org/10.31486/toj.19.0088.

12. Munung NS, Marshall P, Campbell M, et al. Obtaining informed consent for genomics research in Africa: analysis of H3Africa consent documents. *J Med Ethics.* 2016;42(2):132−137. https://doi.org/10.1136/medethics-2015-102796.

13. De Vries J, Munung SN, et al. Regulation of genomic and biobanking research in Africa: a content analysis of ethics guidelines, policies and procedures from 22 African countries. *BMC Med Ethics.* 2017;18.

14. Moodley K, Singh S. It's all about trust: reflections of researchers on the complexity and controversy surrounding biobanking in South Africa. *BMC Med Ethics.* 2016;17.

15. Tindana P, Molyneux CS, Bull S, Parker M. Ethical issues in the export, storage and reuse of human biological samples in biomedical research: perspectives of key stakeholders in Ghana and Kenya. *BMC Med Ethics.* 2014;15(1). https://doi.org/10.1186/1472-6939-15-76.

16. De Vries J, Abayomi A, et al. Addressing ethical issues in H3Africa research−the views of research ethics committee members. *HUGO J.* 2015;9.

17. Ramsay M, De Vries J, Soodyall H, Norris SA, Sankoh O. Ethical issues in genomic research on the African continent: experiences and challenges to ethics review committees. *Hum Genom.* 2014;8(1). https://doi.org/10.1186/s40246-014-0015-x.

18. Nansumba H, Flaviano M, Patrick S, Isaac S, Wassenaar D. Health care users' acceptance of broad consent for storage of biological materials and associated data for research purposes in Uganda. *Wellcome Open Res.* 2022;7:73. https://doi.org/10.12688/wellcomeopenres.17633.1.

19. Tindana P, Molyneux S, Bull S, Parker M. 'It is an entrustment': broad consent for genomic research and biobanks in sub-Saharan Africa. *Develop World Bioeth.* 2019;19(1):9−17. https://doi.org/10.1111/dewb.12178.

20. Amayoa FA, Nakwagala FN, Barugahare J, Munabi IG, Mwaka ES. Understanding of critical elements of informed consent in genomic research: a case of a paediatric HIV-TB research project in Uganda. *Journal of Empirical Research on Human Research Ethics.* 2022;17(4):483−493. https://doi.org/10.1177/15562646221100430.

21. Kengne-Ouafo JA, Millard JD, Nji TM, et al. Understanding of research, genetics and genetic research in a rapid ethical assessment in north west Cameroon. *International Health.* 2016;8(3):197−203. https://doi.org/10.1093/inthealth/ihv034.

22. Traore K, Bull S, Niare A, et al. Understandings of genomic research in developing countries: a qualitative study of the views of MalariaGEN participants in Mali. *BMC Med Ethics.* 2015;16(1). https://doi.org/10.1186/s12910-015-0035-7.

23. Masiye F, Mayosi B, De Vries J. I passed the test! Evidence of diagnostic misconception in the recruitment of population controls for an H3Africa genomic study in Cape Town, South Africa. *BMC Med Ethics.* 2017;18.

24. Mohammed-Ali AI, Gebremeskel EI, Yenshu E, et al. Informed consent in a tuberculosis genetic study in Cameroon: information overload, situational vulnerability and diagnostic misconception. *Res Ethic.* 2022; 18(4):265−280. https://doi.org/10.1177/17470161221106674.

25. Bracken-Roche D, Bell E, Macdonald ME, Racine E. The concept of 'vulnerability' in research ethics: an in-depth analysis of policies and guidelines. *Health Res Pol Syst*. 2017;15(1). https://doi.org/10.1186/s12961-016-0164-6.

26. Campbell MM, Susser E, Mall S, et al. Using iterative learning to improve understanding during the informed consent process in a South African psychiatric genomics study. *PLoS One*. 2017;12(11). https://doi.org/10.1371/journal.pone.0188466.

27. Taiwo RO, Ipadeola J, Yusuf T, et al. Qualitative study of comprehension of heritability in genomics studies among the Yoruba in Nigeria. *BMC Med Ethics*. 2020;21(1). https://doi.org/10.1186/s12910-020-00567-2.

28. Bull S, Farsides B, Ayele FT. Tailoring information provision and consent processes to research contexts: the value of rapid assessments. *J Emp Res n Human Res Ethic*. 2012;7(1):37−52. https://doi.org/10.1525/jer.2012.7.1.37.

29. Gebresilase TT, Deresse Z, Tsegay G, et al. Rapid ethical appraisal: a tool to design a contextualized consent process for a genetic study of podoconiosis in Ethiopia. *Wellcome Open Res*. 2017;2:99. https://doi.org/10.12688/wellcomeopenres.12613.1.

30. Nordling L. A new law was supposed to protect South Africans' privacy. It may block important research instead. *Science*. 2019 https://doi.org/10.1126/science.aax0768.

31. Ball MP, Thakuria JV, Zaranek AW, et al. A public resource facilitating clinical use of genomes. *Proc Natl Acad Sci USA*. 2012;109(30):11920−11927. https://doi.org/10.1073/pnas.1201904109.

32. Beck S, Berner AM, Bignell G, et al. Personal Genome Project UK (PGP-UK): a research and citizen science hybrid project in support of personalized medicine. *BMC Med Genom*. 2018;11(1). https://doi.org/10.1186/s12920-018-0423-1.

33. Krier JB, Kalia SS, Green RC. Genomic sequencing in clinical practice: applications, challenges, and opportunities. *Dialogues Clin Neurosci*. 2016;18(3):299−312. https://doi.org/10.31887/dcns.2016.18.3/jkrier.

34. Middleton A, Milne R, Robarts L, Roberts J, Patch C. Should doctors have a legal duty to warn relatives of their genetic risks? *Lancet*. 2019;394(10215):2133−2135. https://doi.org/10.1016/S0140-6736(19)32941-1.

35. Wolf SM, Branum R, Koenig BA, et al. Returning a research participant's genomic results to relatives: analysis and recommendations. *J Law Med Ethics*. 2015;43(3):440−463. https://doi.org/10.1111/jlme.12288.

36. Healthcare T, South W, London G. Case No: QB-2013-009529. In: *Sussex Partnership NHS Foundation Trust*. 2020. EWHC 455 (QB).

37. Christianson AL. Community genetics in South Africa. *Public Health Genom*. 2001;3(3):128−130. https://doi.org/10.1159/000051122.

38. He LQ, Njambi L, Nyamori JM, et al. Developing clinical cancer genetics services in resource-limited countries: the case of retinoblastoma in Kenya. *Public Health Genom*. 2014;17(4):221−227. https://doi.org/10.1159/000363645.

39. Temtamy SA, Hussen DF. Genetics and genomic medicine in Egypt: steady pace. *Mol Genet Genomic Med*. 2017;5(1):8−14. https://doi.org/10.1002/mgg3.271.

40. Wonkam A, Ngongang Tekendo C, Zambo H, Morris MA. Initiation of prenatal genetic diagnosis of sickle cell anaemia in Cameroon (sub-Saharan Africa). *Prenat Diagn*. 2011;31(12):1210−1212. https://doi.org/10.1002/pd.2896.

41. Wonkam A, Munung NS, Dandara C, Esoh KK, Hanchard NA, Landoure G. Five priorities of African genomics research: the next frontier. *Annu. Rev. Genom. Hum. Genet*. 2022;23:499−521. https://doi.org/10.1146/annurev-genom-111521-102452.

42. Gurdasani D, Carstensen T, Tekola-Ayele F, et al. The African genome variation project shapes medical genetics in Africa. *Nature*. 2015;517(7534):327−332. https://doi.org/10.1038/nature13997.

43. Boulton G. *Science as an Open Enterprise*. The Royal Society; 2012.

44. Pisani E, Whitworth J, Zaba B, Abou-Zahr C. Time for fair trade in research data. *Lancet*. 2010;375(9716): 703−705. https://doi.org/10.1016/S0140-6736(09)61486-0.

45. Vines TH, Albert AYK, Andrew RL, et al. The availability of research data declines rapidly with article age. *Curr Biol*. 2014;24(1):94−97. https://doi.org/10.1016/j.cub.2013.11.014.

46. Bezuidenhout L, Chakauya E. Hidden concerns of sharing research data by low/middle-income country scientists. *Global Bioeth*. 2018;29(1):39−54. https://doi.org/10.1080/11287462.2018.1441780.

47. Mulder N, Adebamowo CA, Adebamowo SN, et al. Genomic research data generation, analysis and sharing − challenges in the African setting. *Data Sci J*. 2017;16. https://doi.org/10.5334/dsj-2017-049.

48. Munung NS, Chi PC, Abayomi A, et al. Perspectives of different stakeholders on data use and management in public health emergencies in sub-Saharan Africa: a meeting report [version 1; peer review: 1 approved, 1 approved with reservations]. *Wellcome Open Res*. 2021;6:1−13. https://doi.org/10.12688/ WELLCOMEOPENRES.16494.1.

49. Hogg A. *Welcome to the Future: Discovery Helping DNA Mapping Move from Sci-Fi to Reality*; 2015 [Online] https://www.biznews.com/innovation/2015/10/14/welcome-to-the-future-how-discovery-helped-dna-mapping-move-from-sci-fi-to-reality. Accessed October 2, 2020.

50. Staunton C, Moodley K. Data mining and biological sample exportation from South Africa: a new wave of bioexploitation under the guise of clinical care? *S Afr Med J*. 2016;106(2):136−138. https://doi.org/10.7196/ SAMJ.2016.v106i2.10248.

51. Munung NS, de Vries J, Pratt B. Towards equitable genomics governance in Africa: guiding principles from theories of global health governance and the African moral theory of Ubuntu. *Bioethics*. 2022;36(4): 411−422. https://doi.org/10.1111/bioe.12995.

52. Blanchard S, Randall I. *South African Scientists Demand the Return of Hundreds of Tribal DNA Samples after a British Institute Was Accused of Trying to Use Them to Make Money*. Mailonline; 2019.

53. Chandrasekharan S, Cook-Deegan R. Gene patents and personalized medicine - what lies ahead? *Genome Med*. 2009;1(9):92. https://doi.org/10.1186/gm92.

54. Sampat B, Williams HL. How do patents affect follow-on innovation? Evidence from the human genome. *Am Econ Rev*. 2019;109(1):203−236. https://doi.org/10.1257/aer.20151398.

55. Smith RD, Thorsteinsdóttir H, Daar AS, Gold ER, Singer PA. Genomics knowledge and equity: a global public health good perspective of the patent system. *Bull World Health Organ*. 2004;82(5):385−389.

56. Emanuel EJ, Wendler D, Grady C. What makes clinical research ethical? *JAMA*. 2000;283(20):2701−2711. https://doi.org/10.1001/jama.283.20.2701.

57. Schroeder D, Ladikas M, Schuklenk U, et al. Sharing the benefits of genetic research. *Br Med J*. 2005; 331(7529):1351−1352. https://doi.org/10.1136/bmj.331.7529.1351.

58. Munung NS, de Vries J. Benefit sharing for human genomics research: awareness and expectations of genomics researchers in sub-saharan Africa. *Ethics and Human Research*. 2020;42(6):14−20. https://doi.org/ 10.1002/eahr.500069.

59. Dauda B, Joffe S. The benefit sharing vision of H3Africa. *Develop World Bioeth*. 2018;18(2):165−170. https://doi.org/10.1111/dewb.12185.

60. Kasule M, Matshaba M, Mwaka E, Wonkam A, De Vries J. Considerations of autonomy in guiding decisions around the feedback of individual genetic research results from genomics research: expectations of and preferences from researchers in Botswana. *Glob Health Epidemiol Genom*. 2022. https://doi.org/10.1155/ 2022/3245206.

61. Ochieng J, Kwagala B, Barugahare J, et al. Perspectives and experiences of researchers regarding feedback of incidental genomic research findings: a qualitative study. *PLoS One*. 2022;17(8):e0273657. https://doi.org/ 10.1371/journal.pone.0273657.

62. Wonkam A, de Vries J. Returning incidental findings in African genomics research. *Nat Genet*. 2020;52(1): 17−20. https://doi.org/10.1038/s41588-019-0542-4.

63. American College of Preventive Medicine (ACPM). *Genetic Testing: A Resource from the American College of Preventive Medicine*. 2010:1–49.

64. Botkin JR, Belmont JW, Berg JS, et al. Points to consider: ethical, legal, and psychosocial implications of genetic testing in children and adolescents. *Am J Hum Genet*. 2015;97(1):6–21. https://doi.org/10.1016/j.ajhg.2015.05.022.

65. American College of Obstetricians and Gynecologists (ACOG). *ACOG Comm Opin*. 2008;410:1–8.

66. Borry P, Stultiens L, Nys H, Cassiman JJ, Dierickx K. Presymptomatic and predictive genetic testing in minors: a systematic review of guidelines and position papers. *Clin Genet*. 2006;70(5):374–381. https://doi.org/10.1111/j.1399-0004.2006.00692.x.

67. Van Noy GE, Genetti CA, McGuire AL, Green RC, Beggs AH, Holm IA. Challenging the current recommendations for carrier testing in children. *Pediatrics*. 2019;143:S27–S32. https://doi.org/10.1542/peds.2018-1099F.

68. Conradie M. *Dysmorphology in the Era of Genomic Medicine*. National Dysmorphology Meeting, Genetic Counsellors; 2023.

69. Knoppers BM, Zawati MH, Sénécal K. Return of genetic testing results in the era of whole-genome sequencing. *Nat Rev Genet*. 2015;16(9):553–559. https://doi.org/10.1038/nrg3960.

70. Auton A, Abecasis GR, Altshuler DM, et al. A global reference for human genetic variation. *Nature*. 2015;526(7571):68–74. https://doi.org/10.1038/nature15393.

71. Altshuler DM, Gibbs RA, Peltonen L, et al. Integrating common and rare genetic variation in diverse human populations. *Nature*. 2010;467(7311):52–58. https://doi.org/10.1038/nature09298.

72. Busby GBJ, Band G, Le QS, et al. Admixture into and within sub-Saharan Africa. *Elife*. 2016;5(2016). https://doi.org/10.7554/elife.15266.

73. Green RC, Berg JS, Grody WW, Kalia SS, Korf BR, Martin CL, American College of Medical Genetics and Genomics. ACMG recommendations for reporting of incidental findings in clinical exome and genome sequencing. *Genet Med*. 2013;15(7):565–574.

74. Nardo-Marino A, Brousse V, Rees D. Emerging therapies in sickle cell disease. *Br J Haematol*. 2020;190(2):149–172. https://doi.org/10.1111/bjh.16504.

75. Ali S, Mumtaz S, Shakir HA, et al. Current status of beta-thalassemia and its treatment strategies. *Mol Genet Genomic Med*. 2021;9(12). https://doi.org/10.1002/mgg3.1788.

76. Piel FB, Patil AP, Howes RE, et al. Global epidemiology of Sickle haemoglobin in neonates: a contemporary geostatistical model-based map and population estimates. *Lancet*. 2013;381(9861):142–151. https://doi.org/10.1016/S0140-6736(12)61229-X.

77. Ma JD, Lee KC, Kuo GM. HLA-B*5701 testing to predict abacavir hypersensitivity. *PLoS Currents*. December 2010. https://doi.org/10.1371/currents.RRN1203.

78. Stainsby CM, Perger TM, Vannappagari V, et al. Abacavir hypersensitivity reaction reporting rates during a decade of HLA-B*5701 screening as a risk-mitigation measure. *Pharmacotherapy*. 2019;39(1):40–54. https://doi.org/10.1002/phar.2196.

79. Mounzer K, Hsu R, Fusco JS, et al. HLA-B*57:01 screening and hypersensitivity reaction to abacavir between 1999 and 2016 in the OPERA® observational database: a cohort study. *AIDS Res Ther*. 2019;16(1). https://doi.org/10.1186/s12981-019-0217-3.

80. La VS, Lefebvre DE, Khalid AF, Decan MR, Godefroy S. Dietary restrictions for people with glucose-6-phosphate dehydrogenase deficiency. *Nutr Rev*. 2019;77(2):96–106. https://doi.org/10.1093/nutrit/nuy053.

81. Lee SWH, Chaiyakunapruk N, Lai NM. What G6PD-deficient individuals should really avoid. *Br J Clin Pharmacol*. 2017;83(1):211–212. https://doi.org/10.1111/bcp.13091.

82. Chu CS, Bancone G, Kelley M, et al. Optimizing G6PD testing for Plasmodium vivax case management: why sex, counseling, and community engagement matter. *Wellcome Open Research*. 2020;5. https://doi.org/10.12688/wellcomeopenres.15700.1.

83. Tayo BO, Kramer H, Salako BL, et al. Genetic variation in APOL1 and MYH9 genes is associated with chronic kidney disease among Nigerians. *Int Urol Nephrol.* 2013;45(2):485–494. https://doi.org/10.1007/s11255-012-0263-4.

84. Ulasi II, Tzur S, Wasser WG, et al. High population frequencies of apol1 risk variants are associated with increased prevalence of non-diabetic chronic kidney disease in the Igbo people from south-eastern Nigeria. *Nephron Clin Pract.* 2013;123(1–2):123–128. https://doi.org/10.1159/000353223.

85. Ekrikpo UE, Mnika K, Effa EE, et al. Association of genetic polymorphisms of TGF-β1, HMOX1, and APOL1 with CKD in Nigerian patients with and without HIV. *Am J Kidney Dis.* 2020;76(1):100–108. https://doi.org/10.1053/j.ajkd.2020.01.006.

86. Freedman BI, Poggio ED. APOL1 genotyping in kidney transplantation: to do or not to do, that is the question? (pro). *Kidney Int.* 2021;100(1):27–30. https://doi.org/10.1016/j.kint.2020.11.025.

87. Mena-Gutierrez AM, Reeves-Daniel AM, Jay CL, Freedman BI. Practical considerations for APOL1 genotyping in the living kidney donor evaluation. *Transplantation.* 2020;104(1):27–32. https://doi.org/10.1097/TP.0000000000002933.

88. Brazas MD, Ouellette BFF, Lewitter F. Continuing education workshops in bioinformatics positively impact research and careers. *PLoS Comput Biol.* 2016;12(6):e1004916. https://doi.org/10.1371/journal.pcbi.1004916.

89. de Klerk M, van der Merwe N, Erasmus J, et al. Incorporating familial risk, lifestyle factors and pharmacogenomic insights into personalised non-communicable disease (NCD) reports for healthcare funder beneficiaries participating in the Open Genome Project. *Annals of Human Genetics (Genetics and Genomics in Health and Disease: A Focus on African Populations).* 2024. In press.

90. Nembaware V, Mulder N. The african genomic medicine training initiative (AGMT): showcasing a community and framework driven genomic medicine training for nurses in Africa. *Front Genet.* 2019;10. https://doi.org/10.3389/fgene.2019.01209.

91. Wonkam A. Are we genetically literate enough for global precision health? *Lancet.* 2023;402(10408):1123–1124. https://doi.org/10.1016/s0140-6736(23)02046-9.

The future of population genomics in developing countries

11

Marlo Möller[1,2], Carene Anne Alene Ndong Sima[1] and Desiree C. Petersen[1]

[1]*South African Medical Research Council Centre for Tuberculosis Research, Division of Molecular Biology and Human Genetics, Faculty of Medicine and Health Sciences, Stellenbosch University, Cape Town, South Africa;* [2]*Centre for Bioinformatics and Computational Biology, Stellenbosch University, Cape Town, South Africa*

Introduction

The past few years have shown a rapid development of new genomic technologies, particularly for whole genome sequencing (WGS). Specifically for human genomics, Francis Collins announced the realization of the genomic era in 2003, marking both the completion of the Human Genome Project and the 50th anniversary of the discovery of the double helix structure of DNA (Fig. 11.1).[2] This was followed by the announcement of personal genomes of well-known geneticists of European descent in 2007,[3,4] with the first African human genome, a Yoruba man from Nigeria, sequenced in 2008.[5] It was only from 2010 that additional African genomes representing various regional populations and human genomes from other developing countries were published.[6] During the so-called postgenomic era, defined as the period after the completion of the human genome project, enormous progress was made.

Not only did a group of omics sciences emerge, including transcriptomics, epigenomics, proteomics, and metabolomics, but a concerted effort was also made to generate large amounts of human DNA sequencing data in population genomics projects such as the 1000 Genomes project, the Human Genome Diversity Project,[7] the Gambian Genome Variation Project,[8] the African Genome Variation Project,[9] the Genetics of Latin American Diversity (GLAD),[10] and others.[11] These population genomics initiatives gave rise to precision medicine and health projects such as the UK Biobank,[12] the All of Us Research Program,[13] and the Trans-Omics for Precision Medicine Program[14,15]—contributing to advancing healthcare, genetics, and research in the respective, mostly developed, countries.[16] All these genetic data further provided the option to perform large-scale genotyping in developing countries as opposed to WGS, which until recently remained very costly. This contributed to an increased number of genome-wide association studies (GWAS) in underrepresented populations, which involved the screening of genetic variants distributed across the entire human genome for trait and disease association. Unfortunately, both off-the-shelf and custom-designed genotyping arrays will ultimately result in future limitations in capturing the vast unknown genetic diversity in developing countries. Based on WGS being now more accessible to many scientists, it can therefore be extrapolated that population

Population Genomics in the Developing World. https://doi.org/10.1016/B978-0-443-18546-5.00011-5

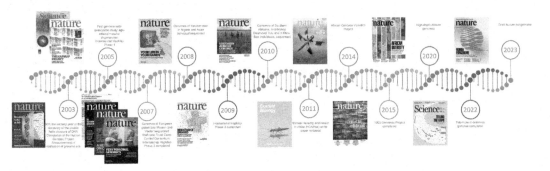

FIGURE 11.1

Genetics and genomics landmarks since the completion of the Human Genome project in 2003.

DNA Timeline Infographic Template by PoweredTemplate. DNA Timeline Infographic; 2020. https://poweredtemplate.com/dna-timeline-infographic-75769/. Accessed 8 July 2023.

genomics can have the same impact in developing countries compared to the developed world, potentially playing a vital role in the evolution of public health genomics in these settings.[17] This chapter will focus on the future of population genomics in developing countries.

Genetic diversity studies and new analysis approaches

Population genetic studies continue to show novel variation within species with extensive genetic diversity observed among human populations—although this diversity is not fully captured by current studies (Fig. 11.2).[18] As mentioned in Chapter 2, developing countries often have rich and diverse populations, with unique evolutionary histories, offering unique opportunities for studying genetic variation. Population genomics can help in understanding the genetic diversity within these populations, leading to improved disease risk assessments, personalized medicine for a more accurate diagnosis and treatment, and targeted interventions that are suitable for settings with limited resources. The question remains, how to increase the number of genetic diversity studies in developing countries to match the efforts of international counterparts that have established large research consortia with access to existing genetic biobanks containing relevant genotype and phenotype information? The answer to this question is multifactorial but is mainly dependent on funding for setting up infrastructure and ensuring that sequencing or other genomic technologies are continuously running at full capacity to screen large sample volumes. The latter has been an ongoing challenge with general resource limitations, including overburdened public healthcare systems, playing a critical role. It often places developing countries in a disadvantaged position where limited genetic analyses can be performed, despite housing most of the worldwide genetic diversity displayed by human populations.

Progress in automated approaches of genetic data may in future circumvent current analysis bottlenecks. One such approach is the automatic extraction of relevant features from genetic data by using neural networks, which implement machine learning as a tool for analysis.[19] This approach has great potential, but it is still a new field with much scope for development. There are several such tools

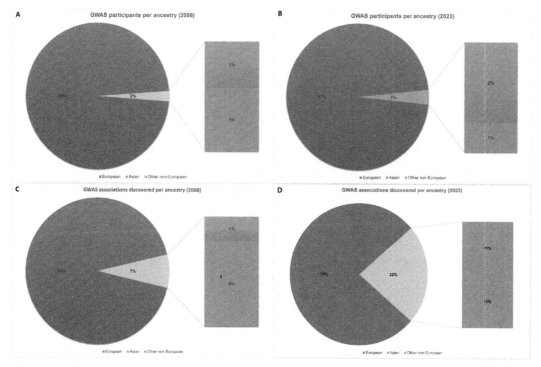

FIGURE 11.2

Genetic studies are not capturing genetic diversity and disease associations across all ancestries, with marginal increased representation between 2008 and 2023. Most GWAS participants are from European populations (A and B), but the inclusion of other ancestries can increase the number of novel GWAS associations (C and D).

Data retrieved from Mills MC, Rahal C. The GWAS Diversity Monitor tracks diversity by disease in real time. Nat Genet. 2020;52(3): 242–243. https://doi.org/10.1038/s41588-020-0580-y.

available for genomics,[20–22] but it was only recently that deep learning frameworks were made available to assist with population genetics.[23–25] One of these tools, named deep neural architectures for DNA (dnadna), is a task-neutral software intended to promote the use, growth, dissemination, and interchange of neural network applications, without requiring a thorough understanding of deep learning or programming (Fig. 11.3).[25]

Downstream of DNA data analysis, highly precise RNA sequencing may now get beyond the problems of errors, artifacts, and bias that have hindered structural biology research, which can also inform population genomics.[26] Investments in the development of protein-prediction algorithms, including AlphaFold (https://alphafold.ebi.ac.uk/) and ESMFold (https://www.cgl.ucsf.edu/chimerax/docs/user/tools/esmfold.html), will soon become evident. The expectation is that the generation of protein structures by these technologies, with precise RNA sequences as the basis, will lead to advancements in biological research, including population genomics.

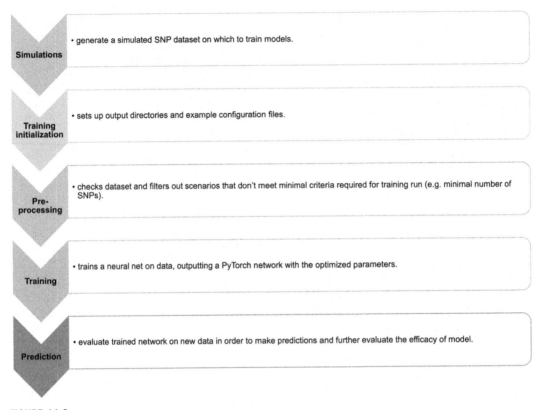

FIGURE 11.3

The five main stages in training a model with dnadna.[25] This involves simulations, training initialization, preprocessing, training, and prediction.

Disease surveillance and prevention

Population genomics can contribute significantly to disease surveillance and prevention efforts in developing countries as demonstrated by the recent coronavirus disease 2019 (COVID-19) pandemic and the response by governments, healthcare workers, and scientists in disease-stricken regions. By studying the genetic factors associated with both communicable and non-communicable diseases prevalent in specific populations, scientific researchers can identify genetic markers, develop predictive models, and implement early interventions for improved public health outcomes. Disease surveillance serves as a vital tool for both the prevention and control of specific diseases endemic to specific developing countries. Determining risk profiles based on population genomics can further provide valuable information regarding the extent of susceptibility and burden of disease across populations within the same regions. The key would be to implement systems that will allow to track the effectiveness of disease surveillance and affordable prevention methods within the public health sector as well as measure the reach within communities that are representative of understudied population groups. Once effective measures are established, it will assist with future interventions to combat any infectious agents or existing heritable diseases that are more common among genetically diverse populations living in developing countries.

Precision medicine and pharmacogenomics

Precision medicine refers to establishing treatment regimens and prevention measures that are effective for a group of individuals with similar genetic, environmental, cultural, and lifestyle factors. Pharmacogenomics is an aspect of precision medicine that involves the study of an individual's response to a particular drug based on the presence of genetic variation, which could be linked to population-specific genetic factors. Developing countries can leverage population genomics to enhance precision medicine initiatives. By studying the genetic makeup of different populations, researchers can identify genetic variants that influence drug response, allowing for tailored treatments and medication dosing based on individual genetic profiles. The implementation of effective precision medicine, or even personalized medicine, interventions is extremely apparent in the developed nations; however, more research is required for developing countries, such as those with unique African populations, where limited genetic data will adversely affect the overall impact of various treatments. The added value of pharmacogenomics could largely benefit understudied populations by identifying trends toward being likely responders to certain drugs or having adverse side effects to specific drugs. Optimizing the drug dosage based on genetic diversity within population groups found within the same geographical location could therefore result in the successful administration of widely tested treatments and long-term approaches for minimizing disease risk within developing countries.

Genomic research consortia

Collaborative efforts between developing countries and international research consortia can accelerate progress in population genomics. Sharing genomic data, resources, and expertise can help in generating comprehensive datasets that encompass diverse populations, enabling better representation and broader generalizability of genomic findings. To date, genetic samples from developing countries—especially those from Africa where the most genetic diversity is present—have not been adequately represented in international consortia. This lack of representation became especially evident during the coronavirus disease 2019 (COVID-19) pandemic that was caused by the severe acute respiratory syndrome coronavirus 2 (SARS-CoV-2).[27] As sizable biobanks were available, developed countries were quickly able to identify human genetic variants linked to COVID-19 severity. However, developing countries such as South Africa—at the forefront of SARS-CoV-2 sequencing efforts[28-30]—were not able to contribute human genetic data as speedily, emphasizing the importance of large-scale genomic initiatives in developing countries.[27]

It has been suggested that each country should have a national plan for genomics showing how the technology would be useful in that specific setting—this would help to highlight the affordability and achievability of these efforts.[31] Recent examples of the successful implementation of large-scale genomic initiatives in developing countries include the Genome of Greece (GoGreece) and DNA

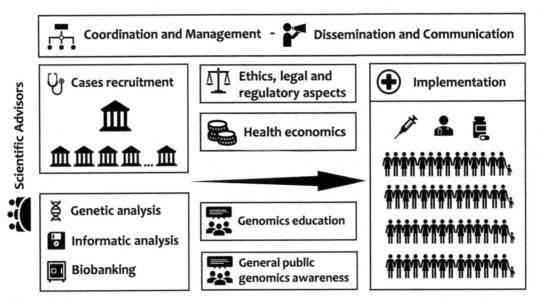

FIGURE 11.4

A model for the implementation of large-scale national genomic projects in developing nations. These steps can ensure successful outcomes.

Reprinted from Patrinos GP, Pasparakis E, Koiliari E, et al. Roadmap for establishing large-scale genomic medicine initiatives in low- and middle-income countries. Am J Hum Genet. 2020;107(4):589–595. https://doi.org/10.1016/j.ajhg.2020.08.005 with permission from Elsevier.

do Brasil (DNABr).[16] The experiences of these initiatives can be used to initiate similar research consortia in other developing settings and bring genomic medicine closer to the bedside (Fig. 11.4). A framework document was compiled that highlighted similar components as identified by GoGreece and DNABr but within the context of Africa.[32] Suggestions were offered on how African governments might work to implement a model for large-scale national genomic projects by leveraging current infrastructure. This will make it possible for countries with limited resources to begin applying effective genomics-based health interventions, drawing on lessons learned from other contexts while making adjustments as needed for the African setting.[32]

The strategy suggested by GoGreece and DNABr is to create a pipeline that initially focuses on case recruitment followed by genetic and bioinformatic studies, underpinned by biobanking.[16] With a limited budget and scarce resources, it would be highly advised that only a subset of participants undergo WGS, while the remainder can be genotyped using microarrays. This would enable researchers to assess the common genomic structure, produce specific pharmacogenomics data, and model disease polygenic scores. This genomic knowledge is easily transferable to first-line interventions in primary care, such as screening campaigns. It is also crucial that a country- or population-specific approach must be taken when choosing the population subgroups to concentrate on, the complex diseases to study, and the technology platforms to be used for implementation.[16] The project could receive the necessary kick-start from crowd-funding and private donations (as in the case of DNABr) or from other research grants (such as GoGreece), which would then persuade other funding bodies, national and possibly regional, to join. This was the case for DNABr, which was included as the population genomics arm in the National Program for Genomics and Precision Medicine of the Brazilian Ministry of Health.[16]

Ethical considerations and data privacy

As population genomics advances, addressing ethical considerations and ensuring data privacy, especially in developing countries, are paramount. Small population groupings in some developing African countries are sufficiently genetically diverse to be easily identified using genetic data. This increases the risk of stigmatization and discrimination for these groups if associated with a particular genetic variant.[32] Even if an individual's status for this variant is unknown, they may be subject to stigmatization based on a perceived elevated risk for all members of their population group of origin if the allele frequency for a deleterious variant is known to be high. It is therefore critical that the populations contributing their genetic diversity to research consortia should benefit from its findings, provided that these benefits are discussed and agreed to during the informed consent process. This is just one of several ethics principles that underlie population genetics, as the unethical use of samples and data can have far-reaching consequences, as was the case when an effort to develop an African microarray for genotyping had to be aborted.[33] More recently, Thermo Fisher Scientific and the family of Henrietta Lacks reached an agreement over the unethical use of cervical cancer cells.[34] The right to refuse medical or scientific experiments without informed consent is enshrined in many constitutions, including the South African constitution, and these rights may only be restricted if doing so is reasonable and justified. For example, COVID-19 impacted on informed consent as individuals were compelled to screening, testing and treatment—sometimes without their consent.[35] African philosophy, which prioritizes the welfare of the community over individual rights, is consistent with this public health strategy.[36] Even so, it was concerning that a recent study in Uganda found that there was substantial therapeutic and/or diagnostic misperception associated with hypothetical genetics and genomics research even in the presence of adequate informed consent processes.[37]

There are also significant regulatory matters that must be considered, particularly in the European setting with the General Data Protection Regulation (GDPR) and European Union In Vitro Diagnostic Regulations acts that may limit data exchange and access to personal information in genomic programs.[16] In the US, state laws function as GDPR equivalents, as federal law has not addressed data security and processing to the extent of the GDPR. As of July 2023, 13 states (California, Colorado, Connecticut, Delaware, District of Columbia, Indiana, Iowa, Montana, Oregon, Virginia, Tennessee, Texas, and Utah) feature some kind of consumer privacy law, while four states are considering similar legislation.[38] In South Africa, the Protection of Personal Information Act has to be considered,[39] Nigeria has a Data Protection Regulation, Mauritius has a Data Protection Act and Kenya announced the Data Protection Bill of 2018.[32] It is not unreasonable to expect that other developing countries will soon implement similar privacy regulations, although different countries may have specific conditions—as is already the case for personal data protection on the African continent (Fig. 11.5). For this reason, informed consent forms should be carefully drafted, especially if genetic data will be returned to participants.[41] Developing countries will need to establish robust ethical frameworks, regulatory guidelines, and data protection mechanisms to safeguard individual privacy while promoting responsible data sharing for research purposes.

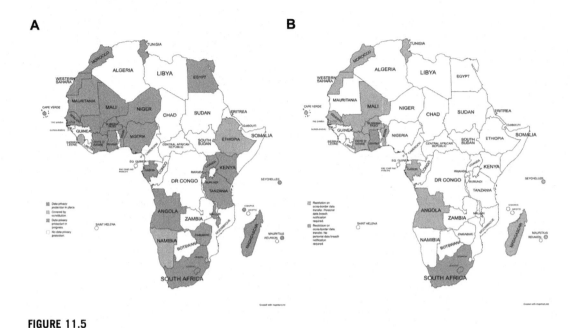

FIGURE 11.5

Maps indicating (A) Personal data protection coverage in Africa and (B) African countries with cross-border data transfer restrictions and personal data breach notification requirements. Different countries may have specific conditions for personal data protection on the African continent.

Data from Deloitte. Privacy Is Paramount: Personal Data Protection in Africa; 2017. (Original work published 2017) and maps created with mapchart.net.

Investing in capacity and infrastructure development

Investing in capacity building and infrastructure development is crucial for the success of population genomics initiatives in developing countries. Building research facilities, training scientists, and establishing collaborations with national and international institutions can empower local researchers to conduct cutting-edge genomic research. Nevertheless, governments and researchers based in developing countries will not have ownership of initiatives or the ability to choose priorities as long as they depend on money from external sources.[42] Furthermore, it is essential for bringing the advantages of genetic medicine to local populations that governments continue to support genomics, as this will allow researchers and medical professionals to continue working in developing settings.[42]

Capacity building efforts for population genomics should focus on training scientists and healthcare professionals in the field, including areas such as data collection, analysis, and interpretation. Workshops, seminars, and collaborative research projects can facilitate knowledge exchange and skill development. Additionally, partnerships between developed and developing countries can play a vital role in knowledge transfer, resource sharing, and technology transfer. By fostering international collaborations, developing countries can leverage the expertise and resources of more advanced nations to accelerate their progress in population genomics, such was the case for the H3Africa initiative.

However, local initiatives have a major role to play, especially in the development of infrastructure—which is equally important.

Adequate laboratory facilities equipped with state-of-the-art sequencing technologies and high-performance computing systems are essential for conducting population genomics research. Investment in robust data storage and management systems is crucial for securely storing and accessing large genomic datasets. Furthermore, establishing ethical guidelines and regulatory frameworks for genomic research ensures the responsible use of data and protects the privacy of individuals. By investing in capacity and infrastructure development, developing countries can overcome barriers and unlock the potential of population genomics to improve healthcare outcomes and address local health challenges. An example of this is the South African Research Infrastructure Roadmap, which led to the establishment of 13 Research Infrastructures. This includes the DIstributed PLatform in OMICS (DIPLOMICS), which aims to strengthen and enable omics capacity and increase access to omics technology and expertise.

Bridging the healthcare gap: public health genomics and precision public health

Population genomics can help to bridge the healthcare gap between developing and developed countries. By improving disease understanding, diagnosis, and treatment options, population genomics can contribute to more equitable healthcare delivery and reduce health disparities globally. However, to be truly egalitarian, population genomics must operate inside a just healthcare system and society and can only contribute if it is integrated with other efforts (economic, political, and social) to address health inequities.[43]

There are still significant disparities between the care provided by public and private healthcare in developing countries. In South Africa, the National Health Insurance (NHI) is a mechanism that has been designed to overcome this challenge.[44] The proposal, which as it stands has attracted much criticism,[45] is to pool funds so that all South Africans can access clinical services based on their needs, regardless of income. This will require robust oversight systems and exemplary leadership but, if implemented correctly, could contribute to the roll out of genetic testing in this country. At present, access to genetic testing is still restricted and newborn screening is not a common hospital procedure in developing countries. A newborn screening program has been implemented in Ghana, and this program benefitted from a genetic-counseling training program initiated by H3Africa investigators.[42]

Conclusion

The future of population genomics in developing countries is very promising, especially with the necessary infrastructure and resources in place (Fig. 11.6). As technology advances continue to occur, the costs associated with reagents decrease, and international collaborations strengthen between continents, developing countries can leverage population genomics to improve their own healthcare outcomes, enhance genetic research that is relevant to local needs, and contribute to global genomic knowledge, particularly for previously understudied regions of the world.

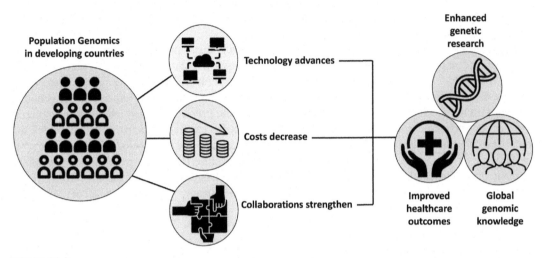

FIGURE 11.6

Population genomics: where we are and where we are heading. As technology advances, costs decrease and collaboration strengthen, we can expect to see enhanced genetics research, improved healthcare outcomes, and a contribution to global genomic knowledge.

References

1. PoweredTemplate. *DNA Timeline Infographic*; 2020. https://poweredtemplate.com/dna-timeline-info-graphic-75769/. Accessed July 8, 2023.
2. Collins FS, Green ED, Guttmacher AE, Guyer MS. A vision for the future of genomics research. *Nature*. 2003;422(6934):835–847. https://doi.org/10.1038/nature01626.
3. Levy S, Sutton G, Ng PC, et al. The diploid genome sequence of an individual human. *PLoS Biol*. 2007;5(10): e254. https://doi.org/10.1371/journal.pbio.0050254.
4. Wheeler DA, Srinivasan M, Egholm M, et al. The complete genome of an individual by massively parallel DNA sequencing. *Nature*. 2008;452(7189):872–876. https://doi.org/10.1038/nature06884.
5. Bentley DR, Balasubramanian S, Swerdlow HP, et al. Accurate whole human genome sequencing using reversible terminator chemistry. *Nature*. 2008;456(7218):53–59. https://doi.org/10.1038/nature07517.
6. Schuster SC, Miller W, Ratan A, et al. Complete Khoisan and Bantu genomes from southern Africa. *Nature*. 2010;463(7283):943–947. https://doi.org/10.1038/nature08795.
7. Bergström A, McCarthy SA, Hui R, et al. Insights into human genetic variation and population history from 929 diverse genomes. *Science*. 2020;367(6484). https://doi.org/10.1126/science.aay5012.
8. Malaria Genomic Epidemiology Network. Insights into malaria susceptibility using genome-wide data on 17,000 individuals from Africa, Asia and Oceania. *Nat Commun*. 2019;10(1):5732. https://doi.org/10.1038/s41467-019-13480-z.
9. Gurdasani D, Carstensen T, Tekola-Ayele F, et al. The African genome variation project shapes medical genetics in Africa. *Nature*. 2015;517(7534):327–332. https://doi.org/10.1038/nature13997.
10. Borda V, Loesch DP, Guo B, et al. Genetics of Latin American Diversity (GLAD) Project: insights into population genetics and association studies in recently admixed groups in the Americas. *bioRxiv*. 2023. https://doi.org/10.1101/2023.01.07.522490.

11. Mallick S, Li H, Lipson M, et al. The Simons genome diversity project: 300 genomes from 142 diverse populations. *Nature*. 2016;538(7624):201−206. https://doi.org/10.1038/nature18964.

12. Downey P, Peakman TC. Design and implementation of a high-throughput biological sample processing facility using modern manufacturing principles. *Int J Epidemiol*. 2008;37(1):i46−i50. https://doi.org/10.1093/ije/dyn031.

13. All of Us Research Hub. https://www.researchallofus.org. Accessed 3 July 2023.

14. Burgess DJ. The TOPMed genomic resource for human health. *Nat Rev Genet*. 2021;22(4):200. https://doi.org/10.1038/s41576-021-00343-x.

15. Taliun D, Harris DN, Kessler MD, et al. Sequencing of 53,831 diverse genomes from the NHLBI TOPMed Program. *Nature*. 2021;590(7845):290−299. https://doi.org/10.1038/s41586-021-03205-y.

16. Patrinos GP, Pasparakis E, Koiliari E, et al. Roadmap for establishing large-scale genomic medicine initiatives in Low- and middle-income countries. *Am J Hum Genet*. 2020;107(4):589−595. https://doi.org/10.1016/j.ajhg.2020.08.005.

17. Molster CM, Bowman FL, Bilkey GA, et al. The evolution of public health genomics: exploring its past, present, and future. *Front Public Health*. 2018;6. https://doi.org/10.3389/fpubh.2018.00247.

18. Mills MC, Rahal C. The GWAS Diversity Monitor tracks diversity by disease in real time. *Nat Genet*. 2020;52(3):242−243. https://doi.org/10.1038/s41588-020-0580-y.

19. Okazaki A, Yamazaki S, Inoue I, Ott J. Population genetics: past, present, and future. *Hum Genet*. 2021;140(2):231−240. https://doi.org/10.1007/s00439-020-02208-5.

20. Kopp W, Monti R, Tamburrini A, Ohler U, Akalin A. Deep learning for genomics using Janggu. *Nat Commun*. 2020;11(1). https://doi.org/10.1038/s41467-020-17155-y.

21. Routhier E, Bin Kamruddin A, Mozziconacci J, Peter R. keras_dna: a wrapper for fast implementation of deep learning models in genomics. *Bioinformatics*. 2021;37(11):1593−1594. https://doi.org/10.1093/bioinformatics/btaa929.

22. Zhang Z, Park CY, Theesfeld CL, Troyanskaya OG. An automated framework for efficiently designing deep convolutional neural networks in genomics. *Nat Mach Intell*. 2021;3(5):392−400. https://doi.org/10.1038/s42256-021-00316-z.

23. Chan J, Spence JP, Mathieson S, Perrone V, Jenkins PA, Song YS. A likelihood-free inference framework for population genetic data using exchangeable neural networks. *Adv Neural Inf Process Syst*; 2018:8594−8605. https://papers.nips.cc/.

24. Flagel L, Brandvain Y, Schrider DR. The unreasonable effectiveness of convolutional neural networks in population genetic inference. *Mol Biol Evol*. 2019;36(2):220−238. https://doi.org/10.1093/molbev/msy224.

25. Sanchez T, Bray EM, Jobic P, et al. dnadna: a deep learning framework for population genetics inference. *Bioinformatics*. 2023;39(1). https://doi.org/10.1093/bioinformatics/btac765.

26. Baptista RP, Kissinger JC, Sheppard DC. Is reliance on an inaccurate genome sequence sabotaging your experiments? *PLoS Pathog*. 2019;15(9):e1007901. https://doi.org/10.1371/journal.ppat.1007901.

27. Petersen DC, Steyl C, Scholtz D, et al. African genetic representation in the context of SARS-CoV-2 infection and COVID-19 severity. *Front Genet*. 2022;13. https://doi.org/10.3389/fgene.2022.909117.

28. Tegally H, Wilkinson E, Giovanetti M, et al. Detection of a SARS-CoV-2 variant of concern in South Africa. *Nature*. 2021;592(7854):438−443. https://doi.org/10.1038/s41586-021-03402-9.

29. Viana R, Moyo S, Amoako DG, et al. Rapid epidemic expansion of the SARS-CoV-2 Omicron variant in southern Africa. *Nature*. 2022;603(7902):679−686. https://doi.org/10.1038/s41586-022-04411-y.

30. Wilkinson E, Giovanetti M, Tegally H, et al. A year of genomic surveillance reveals how the SARS-CoV-2 pandemic unfolded in Africa. *Science*. 2021;374(6566):423−431. https://doi.org/10.1126/science.abj4336.

31. WHO Science Council. Accelerating Access to Genomics for Global Health: Promotion, Implementation, Collaboration, and Ethical, Legal, and Social Issues. https://www.who.int/publications-detail-redirect/9789240052857. Accessed 7 July 2023.

32. Accelerating Excellence in Science in Africa (AESA). *A Framework for the Implementation of Genomic Medicine for Public Health in Africa*; 2021. https://doi.org/10.21955/aasopenres.1115149.1. . Accessed July 4, 2023.

33. Stokstad E. *Major U.K. Genetics Lab Accused of Misusing African DNA*; 2019. https://www.science.org/content/article/major-uk-genetics-lab-accused-misusing-african-dna. Accessed August 7, 2023.

34. Oza A, Lenharo M. How the 'groundbreaking' Henrietta Lacks settlement could change research. *Nature*. 2023. https://doi.org/10.1038/d41586-023-02479-8.

35. McQuoid-Mason DJ. Covid-19 and its impact on informed consent: what should health professionals tell their patients or their proxies? *South African Journal of Bioethics and Law*. 2020;13(1):7−10. https://doi.org/10.7196/SAJBL.2020.v13i1.723.

36. Moodley K. *Bioethics, Medical Law and Human Rights 3/e*. Van Schaik Publishers a division of Media24 Boeke (Pty) Ltd; 2023. https://store.it.si/book/bioethics-medical-law-and-human-rights-3-e/1398661/. Accessed August 7, 2023.

37. Ochieng J, Kwagala B, Barugahare J, Möller M, Moodley K, Cioffi A. Feedback of individual genetic and genomics research results: a qualitative study involving grassroots communities in Uganda. *PLoS One*. 2022; 17(11):e0267375. https://doi.org/10.1371/journal.pone.0267375.

38. US State Privacy Legislation Tracker. 2024. https://iapp.org/resources/article/us-state-privacy-legislation-tracker/. Accessed 7 August 2023.

39. Buys M. Protecting personal information: implications of the protection of personal information (POPI) act for healthcare professionals. *S Afr Med J*. 2017;107(11):954−956. https://doi.org/10.7196/SAMJ.2017.v107i11.12542.

40. Deloitte. *Privacy Is Paramount: Personal Data Protection in Africa*; 2017. https://www2.deloitte.com/za/en/pages/risk/articles/personal-data-protection-in-africa.html. Accessed July 4, 2023.

41. Nembaware V, Johnston K, Diallo AA, et al. A framework for tiered informed consent for health genomic research in Africa. *Nat Genet*. 2019;51(11):1566−1571. https://doi.org/10.1038/s41588-019-0520-x.

42. Lombard Z, Landouré G. Could Africa be the future for genomics research? *Nature*. 2023;614(7946):30−33. https://doi.org/10.1038/d41586-023-00222-x.

43. McGuire AL, Gabriel S, Tishkoff SA, et al. The road ahead in genetics and genomics. *Nat Rev Genet*. 2020; 21(10):581−596. https://doi.org/10.1038/s41576-020-0272-6.

44. Wilkinson M, Gray AL, Wiseman R, et al. Health technology assessment in support of national health insurance in South Africa. *Int J Technol Assess Health Care*. 2022;38(1). https://doi.org/10.1017/S0266462322000265.

45. Prinsen L. *South Africa's National Health Insurance Bill Has Noble Aims but Leaves Too Much Uncertain: It Needs More Work*; 2023. http://theconversation.com/south-africas-national-health-insurance-bill-has-noble-aims-but-leaves-too-much-uncertain-it-needs-more-work-208604. Accessed August 7, 2023.

Index

Note: "Page numbers followed by *f* indicate figures, *t* indicate tables, and *b* indicate boxes."

Printed in the United States
by Baker & Taylor Publisher Services